Satellite Formation Flying

S. Mathavaraj · Radhakant Padhi

Satellite Formation Flying

High Precision Guidance using Optimal and Adaptive Control Techniques

Springer

S. Mathavaraj
U. R. Rao Satellite Center
Indian Space Research Organisation
Bangalore, Karnataka, India

Radhakant Padhi
Department of Aerospace Engineering
Indian Institute of Science
Bangalore, Karnataka, India

ISBN 978-981-15-9630-8 ISBN 978-981-15-9631-5 (eBook)
https://doi.org/10.1007/978-981-15-9631-5

This Springer imprint is published by the registered company Springer Nature Singapore Pte Ltd.
The registered company address is: 152 Beach Road, #21-01/04 Gateway East, Singapore 189721, Singapore

Dedicated to our temple of learning

Indian Institute of Science, Bangalore

www.iisc.ac.in

Foreword by B. N. Suresh

Satellite formation flying is advantageous in facilitating Earth observation, deep space observation, computer tomography of clouds, and many more. To derive maximum benefits from the formation flying of spacecraft, however, it is necessary to guarantee the intended formation with minimum deviation, and that too under the realistic nonlinear scenarios such as elliptic orbits and with J_2 perturbation. To achieve this ambitious objective, *powerful guidance laws based on advanced nonlinear and optimal control theory* are the need of the hour. The authors of this book have discussed *seven such techniques* in this book. They have also summarized their relative merits and demerits and have provided their insightful recommendations under various scenarios. The fundamentals of orbital dynamics as well as the necessary theoretical details of the advanced control concepts have also been elegantly presented, thus enabling the readers to have a better understanding of the ideas without having to go for cross-references.

This book certainly serves as a very good reference for researchers wanting to enter into the fascinating and rapidly evolving field of formation flying of satellites. The nice comparison chapter about various guidance laws, highlighting their potential usage under different conditions, is quite handy to practicing engineers. The accompanying MATLAB program files, which can be downloaded from the publisher's website, are very useful for the readers to quickly experiment. This can enable them to have a better understanding and a 'feel' for the techniques. I am quite impressed by the content of this book. I have no doubt that this book will become a very valuable reference for those who wish to enter and engage themselves in the field of satellite formation flying.

Dr. B. N. Suresh
Chancellor, Indian Institute of Space, Science and Technology, Trivandrum, India

and

Honorary Distinguished Professor, Indian Space Research Organisation, Bangalore, India

and

Past President, Indian National, Academy of Engineering, India

Foreword by R. Pandiyan

Trends in space technology today require small satellites in formation flying to attain specific goals that are otherwise not possible using a single large satellite. In spite of the multitude of research efforts, however, a ready reckoner containing dynamic models followed by advanced control techniques to ensure formation flying was not available. This book bridges this gap and provides an excellent overview and an in-depth study of *suitable dynamic models and modern control techniques to meet the required accuracy*. The gradual portrayal of several methods from LQR and its extensions, the SDRE, the DI technique including its adaptive control augmentation, and finally the powerful MPSP method in order to suitably exploit the inherent nonlinear nature of the satellite formation flying problem is very well documented throughout this book. I wholeheartedly appreciate the authors for bringing out this consolidated and wonderful book in a very lucid and easy-to-understand sequence of chapters. This book will surely provide very valuable insights to those researchers and practicing engineers who work on Satellite Formation Flying.

Dr. R. Pandiyan
Group Director (Retired)
Flight Dynamics Group, URSC
Indian Space Research Organisation
Bangalore, India

and

Former Visiting Faculty
Department of Aerospace Engineering
IIT Madras, Chennai

Foreword by Klaus Schilling

Space exploration is a steady source that offers challenging control engineering problems. Autonomous reaction capabilities were quite essential to handle time-critical situations despite large distances with significant signal propagation delays. Today we see new challenges related to coordinated automation in emerging multi-satellite systems, such as sensor networks in Earth observation and telecommunication networks. In particular, 'Internet of Space' composed of many very small satellites in low Earth orbits promises worldwide data transfer in almost real time. In Earth observation, formations allow from different perspectives a coordinated data acquisition for subsequent sensor data fusion. Thus, satellite formations exhibit extremely high potential for future innovative self-organizing networks in space. It is, therefore, very important for future research to have a solid mathematical basis on the dynamics of formations and appropriate control synthesis approaches to ensure close formation flying of satellites, as presented in this excellent book.

Dr. Klaus Schilling
Professor and Chair for Robotics and Telematics
Julius-Maximilians-University
Würzburg, Germany

and

resident, "Zentrum für Telematik"
Würzburg, Germany

Preface

Satellite formation flying is formally defined as two or more satellites flying in prescribed orbits at an approximately constant separation distance from each other for a given period of time. 'Approximately constant', 'given period of time', etc., are necessary for the definition mainly because the physics of the problem dictates that each satellite must fly in the great circle plane (a plane containing the center of Earth) dictated by its orbital parameters. Hence, unless the satellites are in a circular orbit with fixed angular separation (which happens only in the leading–trailing configuration in the same orbit), constant relative distance between two satellites cannot be maintained continuously. Despite this limitation, an emerging trend across the globe is to have missions involving many inexpensive and distributed satellites flying in formation to achieve common objectives. Some of the key advantages of satellite formation flying include (i) higher redundancy leading to improved fault tolerance (i.e., minimal loss in case of individual failures), (ii) increased mission flexibility and on-orbit reconfiguration, leading to multiple missions within the life span of the satellites, (iii) lower individual launch mass leading to reduced launch cost (small satellites can be launched as co-passengers during heavy satellite launches), and so on.

A variety of interesting and useful applications are possible in satellite formation flying, both as a substitute to single large satellite missions as well as the ones unique to formation flying missions. Wide-area and/or frequent Earth observation for remote sensing and monitoring geographic regions of interest, telescopic formation for solar or deep space observation, computer tomography of clouds for better weather modeling, etc., are a few examples. However, there are several technological challenges to realize such complicated missions using small satellites. Some of the major difficulties include resource constraints (e.g., limited propulsion and battery capacity), limited actuation capability, and hurdles in communication among the individual satellites arising out of limited communication bandwidth as well as attitude perturbations.

It is obvious that good benefits of formation flying can be harvested only if (a) the satellites fly in tight formation with minimum deviation from intended orbits and (b) their attitudes are controlled very well. Topics addressed in this book attempt to address the first issue, i.e., to make sure that the individual satellites fly along their intended orbits with minimum deviation despite the presence of nonlinearity of the relative system dynamics and the major disturbance effect coming from the J_2 perturbation in the gravity model. To achieve this objective, the authors of this book are strongly convinced that the guidance and control algorithm should be based on advanced control theory. However, since the perturbation effects are satellite specific, there is also a need for exploring adaptive control techniques to have customized robust solutions. This is exactly what has been addressed in this book.

Distinct features of this book include the following:

1. Necessary fundamentals of orbital dynamics are included first so that the reader does not have to refer elsewhere for these basic materials. In this chapter, the equation of motion governing the relative motion between two satellites, i.e., the Clohessy–Wiltshire (C–W) equation, has been derived in detail. This is followed by the derivation of the linearized version of it, which is commonly known as Hill's equation. Both of these equations are vital for the concepts discussed in the rest of the book.
2. Seven different techniques for addressing the control of satellite formation flying problem have been discussed in detail in five chapters. Necessary fundamental concepts about these techniques are also included so that the book becomes self-contained as much as possible.
3. An exclusive chapter is included describing the merits and demerits of various techniques presented in the book. This is expected to ease the burden on the practicing engineers to take appropriate decisions about the suitability of a particular technique for their applications.
4. Further, as an additional facility, MATLAB programs are made available on the publisher's webpage for readers to download.

In the following, a brief chapter-wise description is outlined for clear navigation of the book.

In Chap. 2, a brief introduction to the basics of orbital dynamics is provided along with related orbital dynamics terminologies and definitions. First, the equation of motion of two-body problem in the Earth-centered inertial (ECI) frame is derived. Next, using the relative motion between two satellites in the ECI frame, the equation of relative motion in the non-inertial Hill's frame is derived, which is known as the Clohessy–Wiltshire (C–W) equation. The approximate linearized equation of motion of relative satellite dynamics, known as Hill's equation, is presented next. This chapter is concluded by providing details of modeling of perturbation forces on satellites specifically, the effect of oblateness of earth, also called J_2 modeling, on the relative motion of satellites. Even though the thrust realization is not used in this book, a brief discussion about the guidance command realization through thruster actuation is also included in this chapter for the sake of completeness.

In Chap. 3, infinite-time linear quadratic regulator (LQR) theory-based guidance and infinite-time state dependent Riccati Equation (I-SDRE) guidance for satellite formation flying are presented. The linear plant model introduced in the previous chapter is used to synthesize the linear control theory based guidance. Next, two state-dependent coefficient (SDC) formulations of the nonlinear plant model are presented for I-SDRE design. Here, simulation results considering the nonlinear model are included to demonstrate the validity of the LQR and I-SDRE guidance. Note that the invalidity of the LQR for elliptic and/or for large relative distance formation problems is also demonstrated in this chapter.

In Chap. 4, an adaptive LQR approach is presented, where a baseline LQR control theory-based guidance is augmented with two neural networks trained online, resulting in a powerful guidance that can handle nonlinear relative dynamics as well as unwanted external disturbances. One of the two networks approximates the unmodeled dynamics (resulting out of the linearization process) as well as the external J_2 perturbation, which are combined together and treated as state-dependent disturbance terms. The other network augments to the LQR solution to result in a better optimal control theory based guidance for the overall nonlinear system. The significance of this approach is demonstrated using the nonlinear model for elliptic and large relative distance formation problems. Note that owing to the usage of a linear system based LQR controller as the main baseline guidance, this approach may be more acceptable by engineers for implementation in practical missions.

In Chap. 5, an adaptive dynamic inversion (DI) control theory-based guidance is presented. Like in earlier Chap. 4, the baseline nonlinear DI control theory-based guidance is augmented with a neural network. This network trained online captures the disturbances in the system dynamics and hence helps in updating the model in real time. The updated system dynamics is then used in computing the nonlinear control using the DI philosophy. Once again, the overall structure results in an adaptive guidance that is robust to the external disturbance. Simulation results are included in this chapter considering the nonlinear model, J_2 perturbation and correction of large errors in the desired configuration to demonstrate the significance of the proposed approach.

In earlier chapters by infinite-time formulations, the error is usually driven to zero *asymptotically*. However, the satellite formation flying problem is fundamentally a problem where one should ensure formation flying in two neighboring 'orbits' (unlike the formation flying of aerial vehicles). Hence, the right problem formulation should ensure that the relative desired position and velocity vectors are achieved at a *particular time* (not earlier, not later). Once that is achieved, from that time onward, the deputy satellite remains in the desired orbit with respect to the chief satellite. Hence, such a problem formulation should ideally be done under the *finite-time* optimal control paradigm instead.

In Chap. 6, finite-time-based linear quadratic regulator (F-LQR) and state-dependent Riccati Equation (F-SDRE) solution to the nonlinear model is presented. It turns out that the innovative SDC_2 formulation that approximates the nonlinear plant better performs exceptional, which is intuitively obvious. The simulation results from the nonlinear plant are included which corroborate this fact.

In Chap. 7, a finite-time optimal guidance logic is presented for satellite formation flying problem employing the model predictive static programming (MPSP) algorithm. MPSP is a highly computationally efficient nonlinear optimal control solution technique, owing to the necessity of static Lagrange multiplier (instead of dynamic costate variable), recursive computing of the associated sensitivity matrices, etc. MPSP also ensures satisfying hard terminal constraints leading to precision guidance with much lesser terminal errors. A comparative study is presented on the simulation results of MPSP and SDRE guidance techniques. It is shown that the MPSP solution is superior in comparison to the SDRE solution in the sense that it provides minimum terminal state error. Next, another suboptimal guidance logic is presented for satellite formation flying problem using generalized model predictive static programming (G-MPSP) algorithm. The main advantage of G-MPSP over MPSP is that there is no requirement of writing the system dynamics in discrete form to begin with, thereby facilitating higher order numerical methods. This demands relatively more computational power, but results in better accuracy and slightly faster convergence.

In Chap. 8, a comparison study is carried out among four 'prominent guidance techniques' discussed in this book. To be fair to all techniques, the simulation environment is made identical for this comparison study. The advantages and disadvantages of these techniques have been discussed along with the achieved terminal state errors.

Chapter 9 provides a brief summary and discussion of all the topics covered in this book. Finally, a set of conclusions are made and recommendations are provided to facilitate quick selection of a topic in case a reader wants to implement the satellite formation flying problem.

Not much prerequisite is needed to read and understand the contents of the book other than some preliminary fundamental knowledge on modern control theory. However, readers having some prior knowledge in topics such as orbital mechanics, optimal control theory, dynamic inversion control theory, neuro-adaptive control, and so on will have a quick understanding and better appreciation of the topics presented in this book. A reader who needs some basics on the above topics can also listen to some of the NPTEL (National Program on Technology Enhanced Learning) video lectures of the second author, which are available on the NPTEL website[1] as well as in the YouTube platform.

This idea of this book originated when a former master student of the second author, Girish Joshi,[2] completed his Master of Engineering project in 2013. The authors of this book sincerely appreciate and acknowledge his early contributions to the topics discussed in this book. His master project was to experiment with the utility of different advanced control techniques for the satellite formation flying problem and the results were quite promising. After reporting the encouraging results in a

[1]https://nptel.ac.in/courses/101108057/, https://nptel.ac.in/courses/101108047/.

[2] **Girish Joshi** completed his Master of Engineering in 2013 from the Indian Institute of Science, Bangalore under the supervision of the second author and later did his Ph.D. from the University of Illinois at Urbana-Champaign, USA.

couple of conferences (Joshi and Padhi (2013a), Joshi and Padhi (2013b), Maity et al. (2013) Maity, Joshi, and Padhi, Joshi and Padhi (2014), Joshi (2013)), the idea of presenting everything together at one place for the convenience of the readers and users evolved. However, a book needs additional materials, updated results, comparison studies, etc. Due to both time and other constraints, this task lost its priority and got delayed. However, the first author (who is a former Ph.D. student with the second author) willingly took it upon him to do these *additional works* while he was still a Ph.D. student, and thoroughly revised the entire content of this book. This also included a revision of the MATLAB codes, from which the results appearing in this book have been generated. The revised and well-documented codes are now included as part of this book, which can be downloaded from the publisher's website. Moreover, a brief documentation of the program files is provided in the Appendix of this book. With the renewed interest and sustained effort from both the authors for over *2 years*, this book could take the current form.

The authors would like to sincerely acknowledge Dr. R. Pandiyan (former Group Director, Flight Dynamics Group, URSC, ISRO) for his constant encouragement and also for proofreading the manuscript thoroughly. Also, we would like to thank Dr. Ravi Kumar (an ISRO scientist and also a former Ph.D. student of the second author) for sharing his knowledge on the spacecraft thrust realization mechanisms. We also gratefully acknowledge the cover page design effort by Mr. J. Dilip Kumar, which we believe has come out very nice.

Despite the best effort and multiple checks, it is possible that there can be a few unintentional errors in the book. The authors will sincerely appreciate if any such errors (or any other feedback) are conveyed to them.

Bangalore, India

S. Mathavaraj
Radhakant Padhi

References

1. Joshi, G. 2013. Robust and precision satellite formation flying guidance using adaptive optimal control techniques, Ph.D. thesis. https://doi.org/10.13140/RG.2.2.16695.27046.
2. Joshi, G., and R. Padhi. 2013a. Formation flying of small satellites using suboptimal MPSP guidance. In *American Control Conference*, 1584–1589. IEEE.
3. Joshi, G., and R. Padhi. 2013b. Robust satellite formation flying using dynamic inversion with modified state observer. In *IEEE International Conference on Control Applications*, 568–573. IEEE.
4. Joshi, G., and R. Padhi. 2014. Robust satellite formation flying through online trajectory optimization using LQR and neural networks. *IFAC Proceedings Volumes* 47(1): 135–141.
5. Maity, A., G. Joshi, and R. Padhi. 2013. Formation flying of satellites with G-MPSP guidance. In *AIAA Guidance, Navigation, and Control Conference*, 5242.

Contents

About the Authors

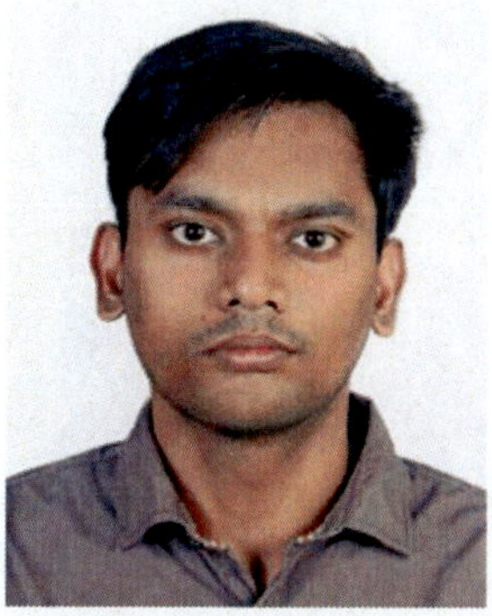

Dr. Mathavaraj S is a space scientist working at the Indian Space Research Organization (ISRO). He earned his Bachelor of Engineering in Aeronautical Engineering from Hindustan College of Engineering, Padur, followed by Master of Technology in Aerospace Engineering from Indian Institute of Technology, Madras and Ph.D. in Aerospace Engineering from Indian Institute of Science, Bengaluru. His interest in space dynamics made him pursue his career in Flight Dynamics Group at U. R. Rao Satellite Center (URSC) of ISRO. His contributions at ISRO include maneuver strategy design for geostationary orbit and lunar capture orbits, attitude profile generation during engine firing and constrained trajectory optimization for soft lunar landing. He is also passionate about formation flying of satellites. For his contribution in the space field, he received ISRO's Young Scientist award in 2018.

Dr. Radhakant Padhi is currently working as a Professor in the Department of Aerospace Engineering and also in the Center for Cyber Physical Systems, Indian Institute of Science (IISc), Bangalore. He earned his Masters and Ph.D. in Aerospace Engineering from IISc Bangalore and Missouri University of Science and Technology, USA respectively. Dr. Padhi is a Fellow of Indian National Academy of Engineers, a Fellow of Institution of Engineers (India) and a Fellow of Institution of Electronics and Telecommunication Engineers. He is an Associate Fellow of American Institute of Aeronautics and Astronautics (AIAA), a Senior Member of Institute of Electrical and Electronics Engineers (IEEE). He is a Vice Chair of the Publication Board of the International Federation of Automatic Control (IFAC) and has been a member of its Council during 2014–2020. Dr. Padhi's research interest lies in synthesis of algorithms in optimal and nonlinear control as well as state estimation. He works on diverse application areas such as control and guidance of aerospace vehicles, biomedical systems, mechanical systems, distributed parameter systems, and industrial process control. He has over 250 publications in international journals and conferences.

1 Introduction and Motivation

Modern-day life is not possible without satellites. Services like cellular phones, emails, internet, television, precision navigation, weather prediction, remote sensing, and so on are easily accessible due to functional satellites. Traditionally, it is done by sending large satellites with enough space for payloads, deployable solar panels, rechargeable power system, and onboard propulsion system. This in general works well, but there are a few important issues. First, large satellites require huge launch vehicles leading to associated problems such as cost, complexity, testbed facility, and so on. Station keeping and attitude maintenance becomes a challenge too since disturbance forces acting on the system (e.g., J_2 perturbation, solar radiation pressure, etc.) are relatively large and one needs to generate equivalent large magnitude counter forces and moments to nullify those disturbance forces. Hence, building and launching of traditional satellites require high-budget missions, which can be quite expensive. Moreover, since any on-orbit failure or malfunction results in huge capital loss, large satellites are usually designed and built to be highly reliable with multiple redundant sub-system hardware. This leads to compromises on the onboard payloads (essentially various sensors to monitor the Earth, sun, stars, etc.), which are very critical to any mission. Because of the complexity, it also leads to a long realization time from design to deployment. Moreover, there is no flexibility in the on-orbit reconfiguration of the mission as the objective of the mission is usually frozen at the inception of the satellite payload, layout, and design. In addition, in case the satellite does not function well, the entire mission fails.

Over the past few decades, there has been immense advancement in the miniaturization of onboard computer, sensor, actuator, and battery technologies. These developments have helped in miniaturing the satellites. Despite this advantage, however, due to their limited size and weight, no meaningful practical mission is possible using small satellites in stand-alone mode. However, in many missions, in principle, one can achieve similar or better performance as compared to a larger satellite using multiple small satellites flying in formation. In view of this, and also because certain missions can only be realized using multiple satellites with some minimum physical separation, an emerging trend across the globe is to have missions involving multiple small, distributed, and inexpensive satellites flying in formation to achieve common objectives.

It is interesting to observe that the definition of spacecraft formation flying is not very precise or universally agreed upon. Most of the space community, however, would agree to the following

S. Mathavaraj and R. Padhi, *Satellite Formation Flying*,
https://doi.org/10.1007/978-981-15-9631-5_1

definition, proposed by NASA's Goddard Space Flight Center: '*Spacecraft formation flying is the maintenance of a desired relative separation, orientation or position between or among the spacecraft flying in a group.*' Note that, unlike aerial vehicles, the desired relative separation between satellites need not remain constant with time to qualify as formation flying. Physics of the problem demands that each satellite must fly in its orbit lying in a respective great circle plane (a plane containing the center of Earth) dictated by its orbital parameters. Hence, unless the satellites are in a circular orbit with fixed angular separation (which happens only in the leading-trailing configuration in the same orbit), constant relative distance between two satellites cannot be maintained continuously. In view of this, satellite formation flying can also be defined as '*Two or more satellites flying in prescribed orbits at an approximately constant separation distance from each other for a given period of time.*' 'Approximately constant', 'given period of time', etc., are necessary in the definition mainly because of the above issue. Even though the relative separation is not maintained constant for all time in satellite formation flying, still there are some key advantages and interesting applications. Some of the key advantages include (i) higher redundancy leading to improved fault tolerance (i.e., minimal loss in case of individual failures), (ii) increased mission flexibility and on-orbit reconfiguration, leading to multiple missions within the life span of the satellites, (iii) lower individual launch mass leading to the reduced launch cost[1] and so on.

A variety of interesting and useful applications are possible in satellite formation flying, both as a substitute to single large satellite missions as well as the ones unique to formation flying missions. There are applications that require distributed systems (where the relative distance among two satellites is not small). Global positioning systems (GPS) is a classic example of such a tightly controlled constellation (a form of formation flying), which has enormous applications in everyday life. A group of satellites in formation flying leads to very long baseline interferometry.[2] Missions such as *CloudCT* [1] are now being envisaged to take collective pictures of a cloud formation region simultaneously, as shown in Fig. 1.1. These images, taken simultaneously from multiple satellites flying in formation, can be used for computed tomographic[3] analysis of a cloud region. This process is expected to lead a better understanding of the cloud formation process, eventually leading to better weather modeling.

In fact, a technology demonstration mission, called *NetSat*, is being pursued by the Zentrum fuer Telematik [2] of Germany in collaboration with Technion, Israel, which comprises of four nano-satellites has been envisaged as shown in Fig. 1.2. The objective is to autonomously control a three-dimensional configuration in space in order to enable new observation methods for climate research as well as for innovative future telecommunication systems [3]. *Telematics earth Observation Mission* (TOM) is another upcoming mission, where the formation of three small satellites for innovative

[1] Reduced mass requires much smaller launch vehicles. However, the launch cost is substantially reduced as small satellites normally fly as pillion rides with conventional prime mission satellites.

[2] An astronomical interferometer is an array of separate telescopes, mirror segments, or radio telescope antennas that work together as a single telescope to provide higher resolution images of astronomical objects such as stars, nebulas, and galaxies by means of interferometry. Interferometry is most widely used in radio astronomy, in which signals from separate radio telescopes are combined. A mathematical signal processing technique called aperture synthesis is used to combine the separate signals to create high-resolution images. In very long baseline interferometry (VLBI), radio telescopes separated by thousands of kilometers are combined to form a radio interferometer with a resolution which would be given by a hypothetical single dish with an aperture thousands of kilometers in diameter. See Wikipedia for more details: https://en.wikipedia.org/wiki/Astronomical_interferometer.

[3] Computer tomography is an imaging procedure that uses computer-processed combinations of many X-ray measurements taken from different angles to produce cross-sectional (tomographic) images (virtual 'slices') of specific areas of a scanned object, allowing the user to see inside the object without cutting.

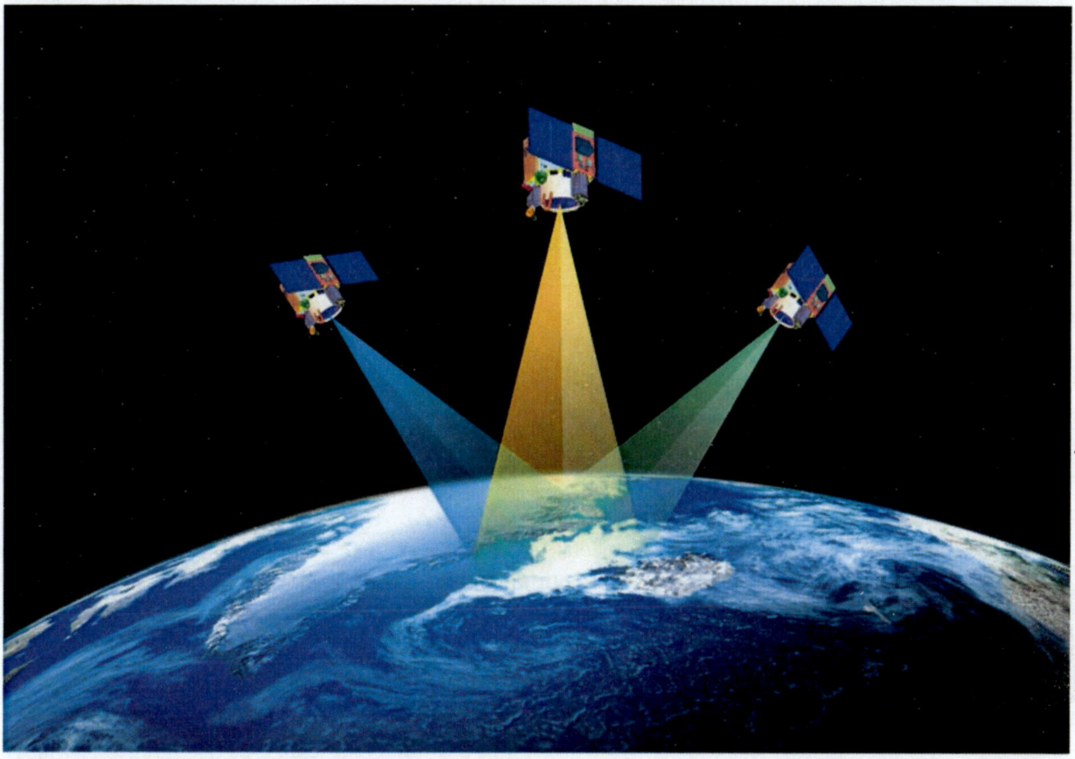

Fig. 1.1 Computed tomography for clouds: a fleet of micro-satellites

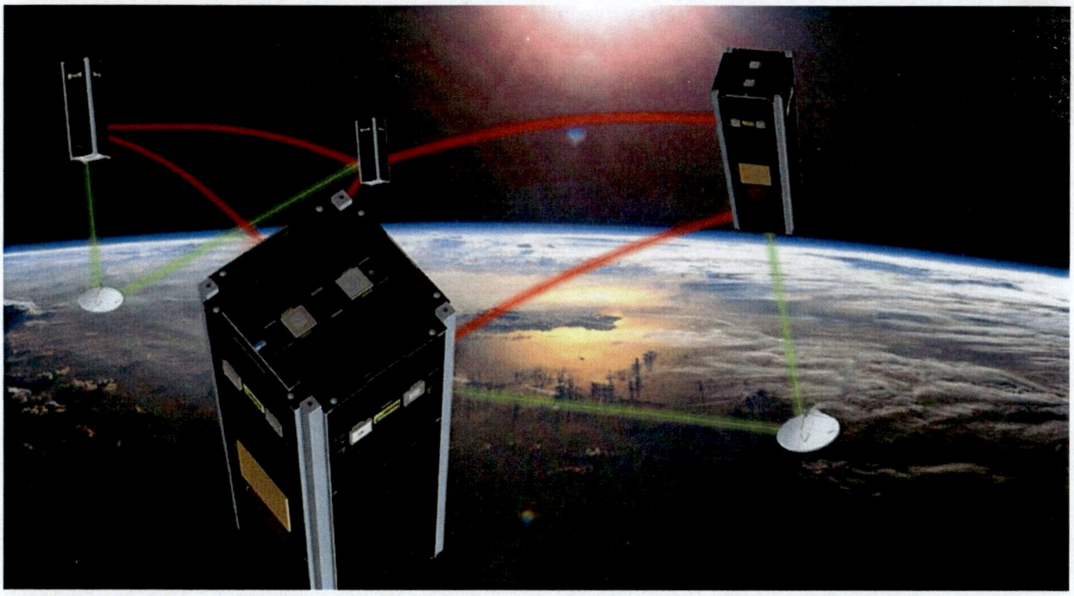

Fig. 1.2 NetSat: networked nano-satellite distributed system control

3D-Earth observation by photogrammetric methods[4] regarding monitoring of volcano eruptions, earthquakes, and environment pollution [4]. However, it can be mentioned here that only a limited number of formation flying missions have been *realized* so far such as GRACE [5], TanDEM-X [6], PRISMA [7] and CanX-4/-5 [8].

The rest of this chapter elaborates on the classification of small satellites, classification of formation flying, and a few other related topics for completeness. It is important that a reader is aware of some of these terminologies and concepts.

1.1 Classification of Small Satellites

One way to classify small satellites is based on the launch mass. Based on this criterion, small satellites are classified into Mini, Micro, Nano, Pico, and Femto satellites, which are elaborated below. However, when the satellite is in orbit, its mass keeps changing due to the usage of propellant mass for orbit and attitude corrections. Therefore, the mass of the fully loaded satellite at the time of launch is taken as a basis for categorizing it into a specific category.

Mini-Satellites: Mini-satellites weigh between 100 and 500 kg. The technologies used in building a mini-satellite are usually borrowed from conventional satellite technologies. The difference can be reduction in the number of payloads (such as transponders or sensors), reduction in overall capability and longevity of the satellite due to reduced onboard fuel and power generation capability (by reducing solar panel size), and so on. These satellites are more or less used in similar applications as regular satellites such as communication, remote sensing, weather monitoring, etc. These are usually equipped with chemical propulsion (hot gas thrusters) for orbit and attitude corrections. Cartosat series, which is used for Indian remote sensing program, launched by Indian Space Research Organisation (ISRO) are examples of this category.

Micro-Satellites Micro-satellites weigh between 10 and 100 kg. Usually, these satellites are used for remote sensing. They are either three-axis-stabilized or spin-axis-stabilized satellites. Microsat series launched by ISRO falls under this category. They are in general utilized for Earth observations during day/night.

Nano-Satellites weigh between 1 and 10 kg range. ExoCube (CP-10) is a space weather nano-satellite developed by California Polytechnic State University. ExoCube's primary mission is to measure the density of hydrogen, oxygen, helium, and nitrogen in the Earth's exosphere.

Pico-Satellites weigh between 0.1 and 1 kg range. They are used in gathering scientific data such as meteorological information, land survey data, marine science, and so on. They also serve purposes such as Earth observation or amateur radio. The CubeSat design, approximately of 1 kg mass, is an example of large pico-satellites. CubeSats are commonly put in orbit by deployers of the International Space Station.

Femto satellites weighs less than 0.1 kg. These satellites can be used in missions such as (i) where a large number of satellites are needed for discrete measurement and sparse sensing, (ii) performance of new material under micro-gravity, etc. Note that the arrival of the nanotechnology, in general, has made it possible to replicate the functionality of an entire satellite on a printed circuit board, which has facilitated the realization of such satellites. For example, the KalamSAT satellite weighing 64 g, designed by a group of Indian students and launched by NASA, falls under this category. The objective of this satellite was to demonstrate the performance of 3D-printed carbon fiber in the micro-gravity environment of space.

[4]Photogrammetry is the science and technology of obtaining reliable information about physical objects and the environment through the process of recording, measuring, and interpreting photographic images and patterns of electromagnetic radiant imagery and other phenomena.

Note that by default small satellites mean mini-satellites. For other categories, one has to explicitly use the appropriate terminology.

1.2 Satellite Formation Flying Architectures

Depending on the configuration, mode of operation, etc., satellite formation flying can be classified into several architectures. The practicing engineers can choose any of these designs based on their problems on-hand.

Trailing configuration: In the trailing configuration, the spacecraft share the same orbit and follow each other on the same path maintaining specified relative angular separation from the center of Earth whenever the chief satellite is at the perigee. Note that in the case of circular orbits, this relative angular separation is maintained constant at all times. However, in case of elliptic orbits, this relative angular separation keeps varying depending on the location of the satellites. Hence it is necessary to define these angles while the chief satellite is at the perigee.

Cluster: In the cluster configuration, a group of satellites are located in formation close to each other and are placed in orbits such that they remain close to each other. Satellites in a cluster usually fly in close formation, but not necessarily in a trailing configuration.

Constellation: A constellation is a group of satellites working in concert. The satellites are more or less uniformly spaced with large separation distances on a globalscale. Constellations normally consist of a set of satellites in an organized orbital plane that covers the entire Earth. Note that the global positioning system (GPS) is the most prominent example of constellation flying.

A brief discussion about the classification of mechanization of guidance schemes for satellite formation flying is included here for completeness. Essentially, formation flying can be achieved by a combination of these approaches, namely either ground-based or autonomous control and centralized or decentralized control. Each approach has its own merits and drawbacks over the other, which are discussed next.

Ground-Based Guidance Versus Autonomous Guidance: In the ground-based guidance, satellite orbital parameters and current conditions are first communicated to ground stations. Utilizing this information, necessary computations are carried out to compute the necessary guidance command. Finally, the guidance command for each satellite is transmitted to the respective orbiting satellite. In autonomous formation flying, on the other hand, the orbital parameters of the current conditions are shared between spacecraft. Subsequently, the necessary computations are performed onboard the satellites to generate the necessary commands of various satellites in formation. The ground-based approach is generally adequate for formation flying when separation distances between the spacecraft are relatively large (at least of the order of a few kilometers). For relatively closer missions, an autonomous approach is preferred.

Centralized Guidance Versus Decentralized Guidance: In the centralized architecture, a central chief satellite does all the necessary computations and transmits the necessary control actions to the other satellites in formation. In small satellite formation flying missions, the centralized architecture is less preferable as the onboard processing power of each satellite is limited. On the other hand, in decentralized architecture, each satellite processes the available information and determines its own guidance commands. This design approach is modular, i.e., adding or removing satellite from the formation does not require any change in the guidance design. It requires minimal communication link between satellites for their relative position update. The addition or loss of a satellite (which is a common scenario in small satellite formation flying) can be accommodated easily which ensures the mission flexibility. Note that only in missions containing a large satellite and several small deputy satellites, the centralized architecture is deployed as that can be the only option.

The theoretical and technological challenges involved in the small satellite formation flying is well documented in [9]. It comprises the details about different types of formation flying configurations as well as the technologies to enable formation flying such as inter-satellite communication, relative navigation, orbit/attitude control, and on-ground test environment requirement for a multi-satellite system.

1.3 Overview on Optimal and Adaptive Control Methods

The techniques used for advanced guidance schemes presented in this book mainly rely on nonlinear optimal and adaptive control techniques. Hence, a brief review of various optimal and adaptive control methods is included in this section for completeness.

1.3.1 Overview of Optimal Control

Optimal control theory [10–12] is a powerful technique for solving many challenging practical trajectory design and guidance problems. It is a field that deals with maximizing or minimizing a cost function subject to applicable constraints, which is done by adjusting the free variables (typically the control variables) optimally. After an appropriate formulation of the problem, for its solution, mainly two approaches are followed in general, namely the indirect approach or the direct approach. In the indirect approach, optimal control problem is formulated using the classical calculus of variations and solved using the classical numerical methods for solving two-point boundary value problems. Traditionally, it leads to computationally intensive iterative procedures for solving it, which lead to 'open-loop control structure'. In direct approach, on the other hand, the optimal control problem is solved by either solving the Hamilton–Jacobi–Bellman (HJB) equation of dynamic programming [12] or by a large-dimensional optimization problem following the transcription (discretization) approach [11].

For a special class of linear quadratic regulator (LQR) problems containing linear (rather linearized) dynamics and a quadratic cost function, a closed-form solution can be derived [13–15]. In fact, for the linear time-invariant systems and infinite-time problems, it leads to a constant gain matrix as well, which can be computed by solving an algebraic Riccati equation offline. Note that, being a nonlinear equation, the Riccati equation offers multiple solutions. One has to necessarily select the positive definite solution which leads to an asymptotically stabilizing optimal controller. Because of its simplicity and ease of implementation, many practicing engineers have applied the LQR control theory based guidance for many aerospace problems, including satellite formation flying [16]. However, since the LQR design is based on the linearized model whereas real-life problems are nonlinear in general, the design does not lead to very good performance. To address this issue partially, researches have attempted to improve the LQR philosophy based design by proposing the so-called 'state-dependent Riccati Equation' (SDRE) method [16–18], which is applicable to a class of control-affine nonlinear systems, where the control variable appears linearly. The idea then is to represent the state-dependent part of the nonlinear system dynamics in the state-dependent coefficient form, after which the nonlinear system dynamics 'appears to be linear'. The cost function is still quadratic, but the weighting matrices are allowed to be functions of the states as well. This leads to an LQR-like problem, but results in a state-dependent Riccati Equation, and hence the approach is named accordingly. Even though finite-time versions of the SDRE approach has been proposed in the literature [19], the most popular version used in the literature is the infinite-time version, which necessitates the solution of an algebraic Riccati equation. However, one should note that the nonlinear Riccati equation now needs to be solved online at each sample time to compute the nonlinear suboptimal solution. The solution is still suboptimal as the costate equation approaches the optimal solution only asymptotically. For a good comprehensive overview of the SDRE approach, a reader is strongly encouraged to see this review paper [20].

It can also be emphasized here that the online solution of the Riccati equation is also not trivial and one can get trapped in the computational issues. Hence, to avoid solving the Riccati equation online, Xin and Balakrishnan have proposed the "θ-D method" [21], which offers an approximate solution of this problem. This method demands the solution of a Riccati equation online and a limited number of Lyapunov equations online. Unlike the Riccati equation, the Lyapunov equation is linear and hence offers a unique and efficient solution. However, it not only suffers from the same drawback of the SDRE approach leading to a suboptimal solution, but in fact it derogates the solution quality a bit further by the way of substituting the original cost function with an approximate cost function. Besides, it also introduces additional tuning parameters, which are not originated from the optimal control theory, to avoid high control magnitude issues. This introduces tuning difficulties as well.

One should note that linear system theory-based formulations such as LQR, SDRE, and θ-D methods are not good candidates for guidance, and in aerospace problems, the system dynamics is invariably nonlinear. Hence, a good guidance scheme should rather attempt to solve a nonlinear trajectory optimization problem. Such problems do not offer closed-form solutions in general. As an alternative, numerical methods such as 'shooting method', 'gradient method', etc. are available [22, 23]. However, these are usually computationally intensive and not suitable for online computations and hence cannot be used for guidance purpose.

As an alternative to the indirect approach, as mentioned before, two direct approaches are also available in the literature, namely the dynamic programming approach and the transcription approach. The dynamic programming approach deals with finding a state feedback form of the optimal control input by solving the Hamilton–Jacobi–Bellman (HJB) equation, which is a nonlinear partial differential equation. The control solution that results from the HJB equation is both necessary and sufficient condition for the optimal cost function and also it is guaranteed to be a stabilizing solution. One of the main limitations of dynamic programming, however, is the fact that the HJB equation is extremely difficult to solve in closed form in general. Attempt to solve it numerically leads to the issue of 'curse-of-dimensionality' where it leads to huge (infeasible) computational and storage requirements. For more details about dynamic programming one can refer to [24–26]. To avoid this computational difficulty, Werbos has proposed '*approximate dynamic programming*' [27] where a discretized approach is followed as in dynamic programming, but results in a set of discretized necessary conditions of optimality as in the indirect approach. Using this result, various versions of the neural network-based 'adaptive-critic' synthesis approaches have also been proposed in the literature, which has been extensively documented in [28]. However, being a neural network-based solution, it suffers from the usual drawbacks such as the necessity of offline training of the neural networks in an expected domain of interest. Moreover, largely these approaches are confined to infinite-time regulation problems since otherwise the relationship between the state and costate becomes dynamic, thereby introducing further complexities.

In yet another direct approach, the transcription philosophy is used where an approximate optimal control problem is formulated by discretizing both the cost function as well as the state equation. The original optimal control problem gets transcribed into this nonlinear programming (static optimization) problem, solution of which leads to the necessary optimal control trajectory (in a discrete sense). Solutions are usually arrived at using the classical numerical solution approaches for solving nonlinear static optimization problems. Even though in principle this looks attractive, one can notice that for improving the accuracy of the solution, smaller sampling time is needed. However, it increases the dimensionality of the optimization problem, thereby leading to computational inefficiency, increased possibility of getting trapped in the local minimum, etc. Some innovations such as 'adaptive grid size', usage of 'sparse algebra', etc., have been introduced to enhance the computational efficiency, but they are usually not adequate to solve the problem online. One can refer to [29–32] for more details about the transcription philosophy in general. Note that standard efficient static optimization solvers are now available such as Sparse Nonlinear OPTimizer (SNOPT) [33] and Interior Point OPTimizer (IPOPT) [34].

A powerful technique that has emerged over the past two decades to solve a class of optimal control problems in near-real time is the model predictive control (MPC) [35–38]. The MPC first uses a dynamic model of the plant to project the system state/output into a finite interval in the future (known as 'prediction horizon'). Subsequently, it minimizes the error between the predicted and desired future state/output vectors to determine control action for the entire horizon. However, only the control action at the current instant is implemented and the process is repeated all over again at the next time instant. This way of implementation makes the control law operate based on feedback philosophy and caters to unwanted effects such as limited modeling imperfection, unwanted disturbance, and so on. It can be mentioned here that the MPC approach follows the 'transcription philosophy'. However, to limit the number of free variables, control action is assumed to be free only for a few instants of time into the future (known as 'control horizon'), after which it is assumed to remain constant for the rest of the prediction horizon. It introduces further sub-optimality, but reduces the dimensionality of the optimization problem thereby resulting in a faster solution. Note that for the nonlinear systems incorporation of nonlinear dynamic models into the MPC formulation results in a nonlinear optimization problem, which in turn leads to a significant increase of computational complexity making it very difficult to implement in real time. Hence applications and success stories of nonlinear MPC is usually rare, even though it has been shown in the literature that nonlinear MPC can lead to significant performance improvement [38, 39]. A good nonlinear MPC algorithm, however, should be computationally very efficient so as to come up with fast solutions without the unwanted approximation, i.e., retaining the control horizon to be the same as the prediction horizon.

An innovative computationally efficient technique, which combines good features of the model predictive control and approximate dynamic programming, named as 'model predictive static programming' (MPSP) has been proposed to solve a class of finite-time optimal control problems with hard terminal constraints [40, 41]. The innovativeness of the MPSP technique lies in successfully converting a dynamic programming problem to a very low-dimensional static programming problem. The main philosophy of the MPSP technique is to exploit the error between the predicted and desired output at the final time and then to rapidly update the entire control history (starting from a guess control history) using a single static Lagrange multiplier so as to minimize this error substantially in the next iteration. Owing to lesser approximations, the method also leads to rapid convergence in general. Several extensions such as Flexible final-time MPSP [42], Generalized MPSP [43], Tracking-oriented MPSP [44], Unscented MPSP [45], etc., have also been proposed to enhance the capability of this innovative design for a wide class of problems.

Before closing this section, it is also worth mentioning here that one of the recently developed transcription approaches of solving the nonlinear optimal control problem is the pseudospectral method [46, 47]. In this method, the state and the control variables are first approximated using spectral basis functions such as Legendre polynomial, Chebyshev polynomial, etc. The idea then is to eventually select these coefficients optimally. In general, this spectral discretization leads to a much smaller dimension of the static optimization problem and hence leads to faster solutions. The static optimization solvers such as SNOPT, IPOPT can be used to obtain the solution. However, one must notice here that even though the solution is substantially faster than the conventional transcription approach, it is not sufficiently fast, in general, to be used for applications that demand online solution (such as guidance of aerospace vehicles). Moreover, it also demands that one should understand more involved mathematical concepts such as quadrature approximation of the cost function, collocation points, etc.

The optimal control theory is very fascinating in general. Several researchers are still continuing their exploration in various aspects of it such as convergence characteristics of numerical algorithms, developing computationally efficient algorithms, proposing closed-form solutions to interesting problems, etc. We believe that new developments of optimal control theory will open up new frontiers and solutions for many practical problems in the future, including satellite formation flying.

1.3.2 Brief Overview of Adaptive Control

It is intuitively obvious that the performance of any model-based control design strongly depends on the model of the system. However, the models used for control design are not reflective of the reality as simplified models are usually preferred for control design. This is done because the high-fidelity models also come with inherent modeling uncertainties, demand additional sensors for information collection, etc. However, it is also intuitively obvious that controllers designed based on simplified models need not perform well and, worse, in some cases may lead to instability as well.

A usual way to handle this issue is to augment the nominal controller with an adaptive controller. The key motivation behind this is to stabilize the system in presence of modeling uncertainties. Broadly two philosophies are followed in the adaptive control domain, namely the direct adaptive control and the indirect adaptive control [48]. The direct adaptive control methods are the ones wherein the controller parameters (i.e., gains) are directly adapted so as to drive the tracking error to zero at the earliest. In the indirect adaptive control, on the other hand, first, the system uncertainties are estimated online and then control parameters are adjusted based on the updated model. Both philosophies rely on some of the fundamental and strong results from nonlinear control such as Lyapunov theory, LaSalle's theorem, Barbalat's Lemma etc. [48, 49]. Each philosophy, however, has its own merit and drawback over the other. It is possible to ensure closed-loop stability of the overall system (i.e., all states) in the indirect adaptive control and hence it is a preferred choice when the time availability to adapt is relatively large. On the other hand, if this is not the case, direct adaptive control is the choice.

A relatively recent technique under the indirect adaptive control is the neuro-adaptive control [50–53], where the neural networks are used to capture the unknown functions resulting out of the modeling uncertainty. The updated model is then used to design the adaptive control. As unknown functions are learned online instead of parameters being identified, this approach usually leads to faster adaptation. More importantly, this approach also makes it fairly independent of the nominal controller. Hence it can be augmented to any nominal controller, thereby making it robust to modeling inaccuracies. In fact, this approach is followed in a few subsequent chapters of this book.

1.3.3 Control-Theoretic Guidance for Satellite Formation Flying

Utilizing the power of advanced control theory, nice guidance laws have been proposed in the literature recently for satellite formation flying. Such a guidance law has been proposed using the linear quadratic regulator [16]. However, the drawback of this guidance is that it is applicable to linear Hill's equation, which is applicable for circular orbits and with close proximity. Hence it is incapable of handling formation flying in elliptic orbits and/or with large separation among satellites. Being a linearized dynamics based formulation, it does not lead to good terminal accuracy. To address this issue, Park et al. [54] have developed a state-dependent Riccati Equation (SDRE) solution, which has been applied for both satellite formation flying (SFF) and station keeping problems. Extending this further, an SDRE-based control technique for non-coplanar formation flying and/or with large separation distance has been developed by Won and Ahn [55]. Successful applicability of this approach has also been demonstrated using numerical simulation studies on a problem where the chief satellite orbits in an elliptic orbit with large eccentricity. In fact, a nice study is available in [56], where comparison studies have been carried out for various linear and nonlinear control technique applied to SFF such as LQR and SDRE. The authors have concluded that the nonlinear control theory-based approach show significant propellant saving over the linear design if the relative orbit to be corrected is large, besides being leading to better terminal accuracy.

It can be mentioned here that a good guidance logic should also compensate for the parameter uncertainties such as mass, thrust misalignment as well as external perturbations such as J_2, solar radiation pressure, aerodynamic forces, etc. Keeping this in mind, Lim et al. [57] have developed a adaptive back-stepping control theory-based guidance for formation flying. This approach handles mass uncertainties and the influence of external perturbation forces. Pongvthithum et al. [58] have developed a universal adaptive control for satellite formation flying for handling time-varying model parameters.

Gurfil et al. [59] have developed an approximate dynamic model inversion-based nonlinear adaptive controller for deep space SFF problem. It can be mentioned here that a popular nonlinear control design technique, in general, is dynamic inversion (DI) method [60], where a linear error dynamics is enforced on the desired output vector in the process of obtaining a nonlinear controller. It turns out that the DI control design technique is quite intuitive and both easy to tune and easy to implement. Note that the DI method is based on the differential geometric concept and relies on the philosophy of feedback linearization in general. An interested reader can find more details about this control design technique in [61]. The major drawback of the DI controller, however, is the fact that it is sensitive to parameter inaccuracy and modeling errors. To address this issue, the neuro-adaptive approach has been proposed in the literature [62, 63]. In general, a Lyapunov-based approach is used to train the neural networks online, which ensures the stability of error dynamics as well as bound on the neural network weights. Moreover, this neural network along with the nominal model is used to invert the model to obtain a robust controller.

In order to handle the practical concerns in SFF missions, researchers have addressed the satellite formation flying problem under the perturbation forces. A detailed control performance study is done on formation flying in presence of gravity perturbation by Sparks [64]. Ahn et al. [65] have developed a robust periodic learning control for trajectory maintenance of satellite formation flying (SFF) under time-periodic influence of external disturbances. Lyapunov-based adaptive nonlinear control law for multi-spacecraft formation flying [66] under the influence of disturbance forces is developed by V. Kapila et al. Also an optimal control based satellite formation guidance under atmospheric drag and J_2 perturbation is developed by Mishne [67]. An interested reader can read these as well as other related references for more details.

1.4 Summary

This chapter primarily introduced the concept and necessity of satellite formation flying mission as well as its several advantages over conventional large satellite counterparts. A brief discussion on the classification of formation flying and satellites involved in such missions are discussed. The various control strategies such as ground-based control and autonomous control are discussed. This chapter motivates the readers about the requirement of control strategy for formation flying of satellites and lays the foundation for further chapters.

References

1. MS Windows NT kernel description. https://cordis.europa.eu/project/id/810370. Accessed 30 Aug 2020.
2. MS Windows NT kernel description. https://www.telematik-zentrum.de/. Accessed 30 Aug 2020.
3. Nogueira, T., J. Scharnagl, S. Kotsiaros, and K. Schilling. 2015. NetSat-4G: A four nano-satellite formation for global geomagnetic gradiometry. In *10th IAA Symposium on Small Satellites for Earth Observation*.
4. MS Windows NT kernel description. https://www.rls-sciences.org/small-satellites.html. Accessed 30 Aug 2020.
5. Davis, E., C. Dunn, R. Stanton, and J. Thomas. 1999. *The GRACE mission: Meeting the technical challenges*. Technical Report, NASA.
6. Huber, S., M. Younis, and G. Krieger. 2010. The TanDEM-X mission: Overview and interferometric performance. *International Journal of Microwave and Wireless Technologies* 2 (3–4): 379–389.
7. Gill, E., S. D'Amico, and O. Montenbruck. 2007. Autonomous formation flying for the PRISMA mission. *Journal of Spacecraft and Rockets* 44 (3): 671–681.
8. Kahr, E., N. Roth, O. Montenbruck, B. Risi, and R.E. Zee. 2018. GPS relative navigation for the CanX-4 and CanX-5 formation-flying nanosatellites. *Journal of Spacecraft and Rockets* 55 (6): 1545–1558.
9. Schilling, K. 2020. I-3d: Formations of small satellites. In *Nanosatellites: Space and ground technologies, operations and economics*, 327–339.
10. Roberts, S.M., and J.S. Shipman. 1972. *Two point boundary value problems: Shooting methods*. New York, USA: American Elsevier Publishing Company Inc.
11. Betts, J.T. 2001. *Practical methods for optimal control using nonlinear programming*. Philadelphia, USA: The Society for Industrial and Applied Mathematics.
12. Bryson, A.E., and Y.C. Ho. 1975. *Applied optimal control*. Hemisphere Publishing Corporation.
13. Anderson, B.D., and J.B. Moore. 2007. *Optimal control: Linear quadratic methods*. Courier Corporation.
14. Naidu, D.S. 2002. *Optimal control systems*. CRC Press.
15. Sinha, A. 2007. *Linear systems: Optimal and robust control*. CRC Press.
16. Jin, X., and H. Lifu. 2011. *Formation keeping of micro-satellites LQR control algorithms analysis*, vol. 4.
17. Wernli, A., and G. Cook. 1975. Suboptimal control for the nonlinear quadratic regulator problem. *Automatica* 11 (1): 75–84.
18. Cloutier, J.R. 1997. State-dependent Riccati equation techniques: An overview. In *Proceedings of the 1997 American Control Conference*, vol. 2, 932–936. IEEE.
19. Heydari, A., and S.N. Balakrishnan. 2012. Approximate closed-form solutions to finite-horizon optimal control of nonlinear systems. In *American Control Conference (ACC)*, 2657–2662. IEEE.
20. Cimen, T. 2012. Survey of state-dependent Riccati equation in nonlinear optimal feedback control synthesis. *Journal of Guidance, Control, and Dynamics* 35 (4): 1025–1047.
21. Xin, M., and S. Balakrishnan. 2005. A new method for suboptimal control of a class of non-linear systems. *Optimal Control Applications and Methods* 26 (2): 55–83.
22. Osborne, M.R. 1969. On shooting methods for boundary value problems. *Journal of Mathematical Analysis and Applications* 27 (2): 417–433.
23. Keller, H.B. 2018. *Numerical methods for two-point boundary-value problems*. Courier Dover Publications.
24. Bellman, R., and R.E. Kalaba. 1965. *Dynamic programming and modern control theory*, vol. 81. Citeseer.
25. Larson, R.E., J.L. Casti, and J.L. Casti. 1978. *Principles of dynamic programming*. New York: Marcel Dekker.
26. Bertsekas, D.P. 1995. *Dynamic programming and optimal control*, vol. 1. MA: Athena scientific Belmont.
27. Werbos, P.J. 1992. *Approximate dynamic programming for real-time control and neural modeling*, ed. D.A. White, D.A. Sofge. New York: Van Nostrand Reinhold.
28. Powell, W.B. 2004. *Handbook of learning and approximate dynamic programming*, vol. 2. Wiley.
29. Bock, H.G., and K.-J. Plitt. 1984. A multiple shooting algorithm for direct solution of optimal control problems. *IFAC Proceedings Volumes* 17 (2): 1603–1608.
30. Betts, J.T. 1994. Issues in the direct transcription of optimal control problems to sparse nonlinear programs. In *Computational optimal control*, 3–17. Springer.
31. Bulirsch, R., and D. Kraft. 2012. *Computational optimal control*, Vol. 115. Birkhäuser.
32. Subchan, S., and R. Zbikowski. 2009. *Computational optimal control*. Dhrivenham: Cranfield University.
33. Gill, P.E., W. Murray, and M.A. Saunders. 2005. SNOPT: An SQP algorithm for large-scale constrained optimization. *SIAM Review* 47 (1): 99–131.
34. Kawajir, Y., C. Laird, and A. Wachter. 2006. *Introduction to IPOPT: A tutorial for downloading, installing, and using IPOPT*. Technical Report, November, 2006.
35. Mayne, D.Q., J.B. Rawlings, C.V. Rao, and P.O. Scokaert. 2000. Constrained model predictive control: Stability and optimality. *Automatica* 36 (6): 789–814.
36. Pluymers, B., J. Rossiter, J. Suykens, and B. De Moor. 2005. A simple algorithm for robust MPC. In *Proceedings of the IFAC World Congress*.

37. Fernandez-Camacho, E., and C. Bordons-Alba. 1995. *Model predictive control in the process industry*. Springer.
38. Allgöwer, F., and A. Zheng. 2012. *Nonlinear model predictive control*, vol. 26. Birkhäuser.
39. Qin, S.J., and T.A. Badgwell. 2000. An overview of nonlinear model predictive control applications. In *Nonlinear model predictive control*, 369–392. Springer.
40. Padhi, R., and M. Kothari. 2009. Model predictive static programming: A computaionally efficient technique for suboptimal control design. *International Journal of Innovative Computing, Information and Control* 5 (2): 399–411.
41. Halbe, O., R.G. Raja, and R. Padhi. 2013. Robust reentry guidance of a reusable launch vehicle using model predictive static programming. *Journal of Guidance, Control, and Dynamics* 37 (1): 134–148.
42. Maity, A., R. Padhi, S. Mallaram, and M. Manickavasagam. 2012. MPSP guidance of a solid motor propelled launch vehicle for a hypersonic mission. In *AIAA Guidance, Navigation, and Control Conference*, 4474.
43. Maity, A., H.B. Oza, and R. Padhi. 2014. Generalized model predictive static programming and angle-constrained guidance of air-to-ground missiles. *Journal of Guidance, Control, and Dynamics* 37 (6): 1897–1913.
44. Kumar, P., B.B. Anoohya, and R. Padhi. 2019. Model predictive static programming for optimal command tracking: A fast model predictive control paradigm. *Journal of Dynamic Systems, Measurement, and Control* 141 (2): 021014.
45. Mathavaraj, S., and R. Padhi. 2019a. Unscented MPSP for optimal control of a class of uncertain nonlinear dynamic systems. *Journal of Dynamic Systems, Measurement, and Control* 141 (6): 065001.
46. Ross, I.M., and F. Fahroo. 2002. A direct method for solving nonsmooth optimal control problems. *IFAC Proceedings Volumes* 35 (1): 479–484.
47. Elnagar, G.N., and M.A. Kazemi. 1998. Pseudospectral Chebyshev optimal control of constrained nonlinear dynamical systems. *Computational Optimization and Applications* 11 (2): 195–217.
48. Krstic, M., I. Kanellakopoulos, and P. Kokotovic. 1995. *Nonlinear and adaptive control design*. New York: Willey.
49. Egardt, B. 1979. *Stability of adaptive controllers*, vol. 20. Springer.
50. Moody, J.E. 1992. The effective number of parameters: An analysis of generalization and regularization in nonlinear learning systems. In *Advances in neural information processing systems*, 847–854.
51. Padhi, R., N. Unnikrishnan, and S.N. Balakrishnan. 2007. Model-following neuro-adaptive control design for non-square, non-affine nonlinear systems. *IET Control Theory Applications* 1 (6): 1650–1661.
52. Suykens, J.A., J.P. Vandewalle, and B.L. de Moor. 2012. *Artificial neural networks for modelling and control of non-linear systems*. Springer Science & Business Media.
53. Ge, S., C. Hang, T. Lee, and T. Zhang. 2013. *Stable adaptive neural network control*. The International Series on Asian Studies in Computer and Information Science. US: Springer.
54. Park, H.E., S.Y. Park, and K.H. Choi. 2011. Satellite formation reconfiguration and station keeping using SDRE technique. *Aerospace Science And Technology* 15: 440–452.
55. Won, C.H., and H.S. Ahn. 2003. Nonlinear orbital dynamic equations and state-dependent Riccati equation control of formation flying satellites. *Journal of the Astronautical Sciences* 51 (4): 433–450.
56. Irvin, D.J., and D.R. Jacques. 2002. A study of linear versus nonlinear control techniques for the reconfiguration of satellite formations. In *Advances in the astronautical sciences* 589–608.
57. Lim, H., H. Bang, and S. Lee. 2006. Adaptive backstepping control for satellite formation flying with mass uncertainty. *Journal of Astronomy and Space Sciences* 23 (4): 405.
58. Pongvthithum, R., S. Veres, S. Gabriel, and E. Rogers. 2005. Universal adaptive control of satellite formation flying. *International Journal of Control* 78 (1): 45–52.
59. Gurfil, P., M. Idan, and N.J. Kasdin. 2003. Adaptive neural control of deep-space formation flying. *Journal of Guidance, Control, and Dynamics* 26 (3): 491–501.
60. Enns, D., D. Bugajski, R. Hendrick, and G. Stein. 1994. Dynamic inversion: An evolving methodology for flight control design. *International Journal of Control* 59 (1): 71–91.
61. Slotine, J.J., and W. Li. 1991. *Applied nonlinear control*. Prentice Hall.
62. Kim, B.S., and A.J. Calise. 1997. Nonlinear flight control using neural networks. *AIAA Journal of Guidance, Control and Dynamics* 20 (1): 26–33.
63. Padhi, R., S.N. Balakrishnan, and N. Unnikrishnan. 2007. Model-following neuro-adaptive control design for non-square, non-affine nonlinear systems. *IET Control Theory Application* 1 (6): 1650–1661.
64. Sparks, A. 2000. Satellite formation keeping control in the presence of gravity perturbations. *Proceedings of American Control Conference* 2: 844–848.
65. Ahn, H.S., K.L. Moore, and Y.Q. Chen. 2010. Trajectory keeping in satellite formation flying via robust periodic control. *International Journal of Robust and Non-linear Control* 20: 1655–1666.
66. De Queiroz, M., V. Kapila, and Q. Yan. 2000. Adaptive nonlinear control of multiple spacecraft formation flying. *Journal of Guidance, Control, and Dynamics* 23 (3): 385–390.
67. Mishne, D. 2004. Formation control of satellites subject to drag variations and J_2 perturbations. *Journal of Guidance, Control, and Dynamics* 27 (4): 685–692.

Satellite Orbital Dynamics

2

Orbital dynamics is primarily concerned with the motion of orbiting celestial and man-made bodies. A well-studied specific orbital dynamics problem is the classic *two-body problem*, where two celestial bodies keep moving under the gravitational influence of each other. This chapter presents an overview of the two-body orbital mechanics first. Satellite orbital dynamics is presented next, which is a special case of the two-body problem, where the mass of one celestial body (e.g., the satellite) is negligible as compared to the body around which it orbits. Subsequently, the relative motion of two satellites is presented, which is used to synthesize the guidance schemes for formation flying.

2.1 The Keplerian Two-body Problem

Figure 2.1 shows the two celestial bodies B_1 and B_2 with masses m_1 and m_2, respectively, moving under the gravitational influence of each other in the inertial frame of reference $(\mathbf{e}_{\hat{x}}, \mathbf{e}_{\hat{y}}, \mathbf{e}_{\hat{z}})$. The origin O of the inertial frame remain stationary or may move with constant velocity (relative to the fixed stars), but the axes $(\mathbf{e}_{\hat{x}}, \mathbf{e}_{\hat{y}}, \mathbf{e}_{\hat{z}})$ do not rotate. $\mathbf{R}_1$ and $\mathbf{R}_2$ represent the position vectors of the center of masses of the two bodies in the inertial frame of reference with $(\hat{x}_1, \hat{y}_1, \hat{z}_1)$ and $(\hat{x}_2, \hat{y}_2, \hat{z}_2)$ being their respective position coordinates. Hence $\mathbf{R}_1$ and $\mathbf{R}_2$ can be written as

$$\begin{aligned} \mathbf{R}_1 &= \hat{x}_1\mathbf{e}_{\hat{x}} + \hat{y}_1\mathbf{e}_{\hat{y}} + \hat{z}_1\mathbf{e}_{\hat{z}} \\ \mathbf{R}_2 &= \hat{x}_2\mathbf{e}_{\hat{x}} + \hat{y}_2\mathbf{e}_{\hat{y}} + \hat{z}_2\mathbf{e}_{\hat{z}} \end{aligned} \tag{2.1}$$

Next, the relative position vector of B_2 with respect to B_1 is defined as $\mathbf{r} = \mathbf{R}_2 - \mathbf{R}_1$. Using Eq. (2.1), relative position vector $\mathbf{r}$ is written as follows:

$$\mathbf{r} = \left(\hat{x}_2 - \hat{x}_1\right)\mathbf{e}_{\hat{x}} + \left(\hat{y}_2 - \hat{y}_1\right)\mathbf{e}_{\hat{y}} + \left(\hat{z}_2 - \hat{z}_1\right)\mathbf{e}_{\hat{z}} \tag{2.2}$$

The body B_1 is acted upon by the gravitational pull from body B_2. This gravitational force of attraction $\mathbf{F}_{12}$ acts along the line joining the center of the two bodies along the direction from B_1 to B_2. The unit vector $\mathbf{e}_r$ along this direction can be defined as

S. Mathavaraj and R. Padhi, *Satellite Formation Flying*,
https://doi.org/10.1007/978-981-15-9631-5_2

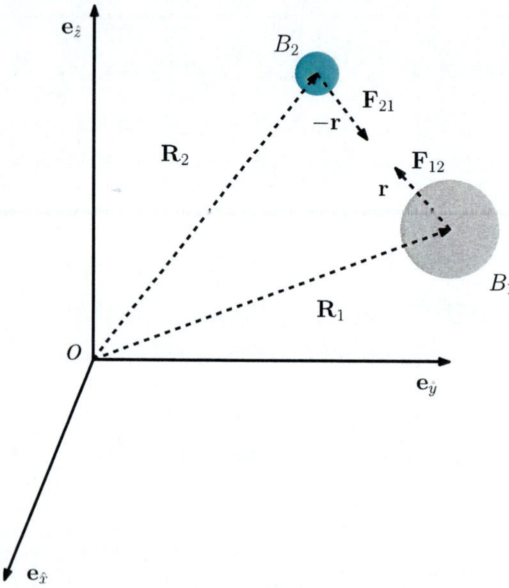

Fig. 2.1 Two-body problem in inertial frame

$$\mathbf{e}_r = \frac{\mathbf{r}}{|\mathbf{r}|} \tag{2.3}$$

Using Newton's law of universal gravitation, which assumes the masses to be concentrated at the respective center of masses of the two bodies, the gravitational force that acts on B_1 by B_2 can be written as

$$\mathbf{F}_{12} = \frac{Gm_1m_2}{r^2}\mathbf{e}_r \tag{2.4}$$

where $G = 6.672 \times 10^{-11} \frac{\text{N m}^2}{\text{kg}^2}$ is the universal Gravitational constant. Similarly, the gravitational force acting on B_2 by B_1 can be written as

$$\mathbf{F}_{21} = \frac{Gm_1m_2}{r^2}(-\mathbf{e}_r) \tag{2.5}$$

which conforms to Newton's third law of motion (i.e., the action and reaction forces are equal and opposite).

One should notice, however, that gravitational force is not the only force that acts on the celestial objects. There can be additional perturbation forces such as distributed effect of gravitational force (which depends on the mass distribution of the interacting bodies), gravitational attraction from other bodies, aerodynamic forces (if applicable) and so on. In man-made objects (such as artificial satellites) it can also contain the 'control force', which can be intelligently manipulated to achieve certain mission-specific desired objectives. Accounting for these realities and using Newton's second law of motion, the absolute force on body B_1 in the inertial frame of reference can be written as

$$\mathscr{F}_{12} = \mathbf{F}_{12} + \mathbf{F}_{1c} + \mathbf{F}_{1p} = m_1\ddot{\mathbf{R}}_1 \tag{2.6}$$

where $\mathbf{F}_{1c}$ is the control force and $\mathbf{F}_{1p}$ is the perturbation force. Using Eqs. (2.4)–(2.6), the absolute force can be rewritten as

$$m_1\ddot{\mathbf{R}}_1 = \frac{Gm_1m_2}{r^2}\mathbf{e}_r + \mathbf{F}_{1c} + \mathbf{F}_{1p} \tag{2.7}$$

Similarly, writing the above set of equation for body B_2 one obtains

$$m_2\ddot{\mathbf{R}}_2 = -\frac{Gm_1m_2}{r^2}\mathbf{e}_r + \mathbf{F}_{2c} + \mathbf{F}_{2p} \tag{2.8}$$

Dividing both sides of Eq. (2.7) by m_1 and both sides of Eq. (2.8) by m_2, these equations are rewritten as follows:

$$\ddot{\mathbf{R}}_1 = \frac{Gm_2}{r^2}\mathbf{e}_r + \frac{\mathbf{F}_{1c}}{m_1} + \frac{\mathbf{F}_{1p}}{m_1} \tag{2.9}$$

$$\ddot{\mathbf{R}}_2 = -\frac{Gm_1}{r^2}\mathbf{e}_r + \frac{\mathbf{F}_{2c}}{m_2} + \frac{\mathbf{F}_{2p}}{m_2} \tag{2.10}$$

Using Eqs. (2.9) and (2.10), the relative acceleration between the bodies B_1 and B_2 can be written as

$$\ddot{\mathbf{r}} = \ddot{\mathbf{R}}_2 - \ddot{\mathbf{R}}_1 = -\frac{Gm_1}{r^2}\mathbf{e}_r + \frac{\mathbf{F}_{2c}}{m_2} + \frac{\mathbf{F}_{2p}}{m_2} - \frac{Gm_2}{r^2}\mathbf{e}_r - \frac{\mathbf{F}_{1c}}{m_1} - \frac{\mathbf{F}_{1p}}{m_1} \tag{2.11}$$

Rewriting the above equation by combining the terms results in

$$\ddot{\mathbf{r}} = -\frac{G(m_1+m_2)}{r^2}\mathbf{e}_r + \frac{\mathbf{F}_{2c}}{m_2} + \frac{\mathbf{F}_{2p}}{m_2} - \frac{\mathbf{F}_{1c}}{m_1} - \frac{\mathbf{F}_{1p}}{m_1} \tag{2.12}$$

The center of mass C of the two bodies B_1 and B_2 from the origin of the inertial frame (see Fig. 2.2) is calculated as

$$\mathbf{R}_C = \frac{(m_1\mathbf{R}_1 + m_2\mathbf{R}_2)}{m_1 + m_2} \tag{2.13}$$

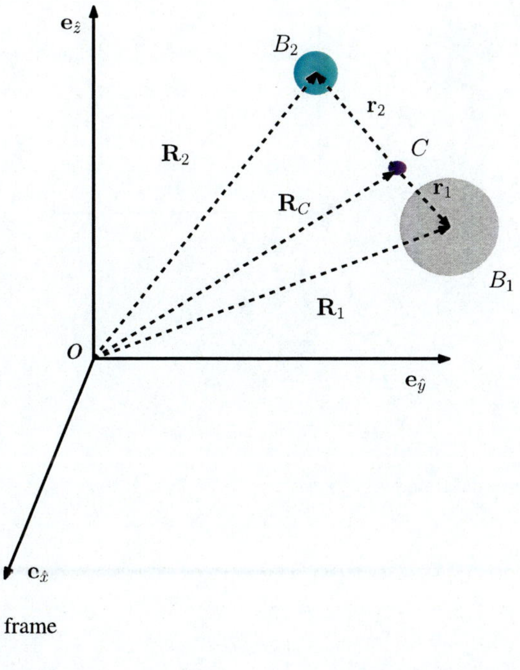

Fig. 2.2 Barycenter in inertial frame

Therefore, the absolute velocity and absolute acceleration of C are

$$\dot{\mathbf{R}}_C = \frac{(m_1\dot{\mathbf{R}}_1 + m_2\dot{\mathbf{R}}_2)}{m_1 + m_2} \tag{2.14}$$

$$\ddot{\mathbf{R}}_C = \frac{(m_1\ddot{\mathbf{R}}_1 + m_2\ddot{\mathbf{R}}_2)}{m_1 + m_2} \tag{2.15}$$

If control and perturbation force is assumed to be negligible, then it is clear from Eqs. (2.7) and (2.8) that

$$m_1\ddot{\mathbf{R}}_1 + m_2\ddot{\mathbf{R}}_2 = 0 \tag{2.16}$$

It means $\ddot{\mathbf{R}}_C = 0$, i.e., there is no acceleration acting at the center of mass C of two bodies B_1 and B_2. This center of mass C is known as the *Barycenter*. Since it is a non-accelerating point, it can also be treated as the origin of the inertial frame, which is usually the case.

Next, let us consider this frame of reference (see Fig. 2.3), where the origin O is the barycenter of the two-body system. Defining r_1 as the distance between the center of B_1 and C; r_2 as the distance between the center of B_2 and C as shown in the Fig. 2.2. Then, Eq. (2.13) can be written as

$$\mathbf{0} = \frac{(m_1(-\mathbf{r}_1) + m_2\mathbf{r}_2)}{m_1 + m_2} \tag{2.17}$$

which in turn implies

$$\mathbf{r}_1 = \frac{m_2}{m_1}\mathbf{r}_2 \tag{2.18}$$

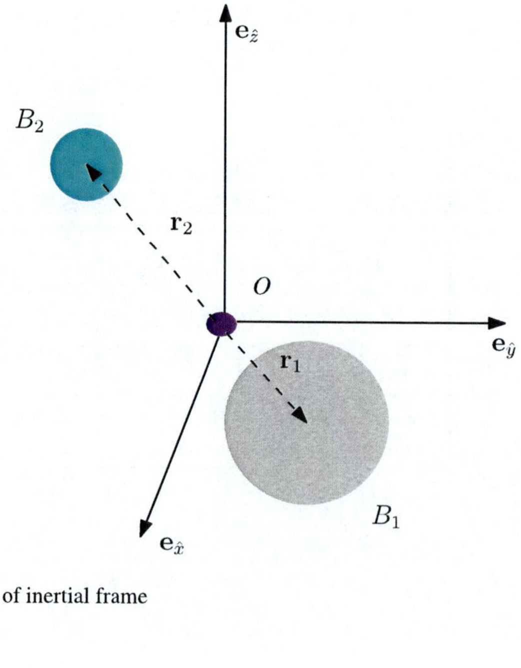

Fig. 2.3 Barycenter as origin of inertial frame

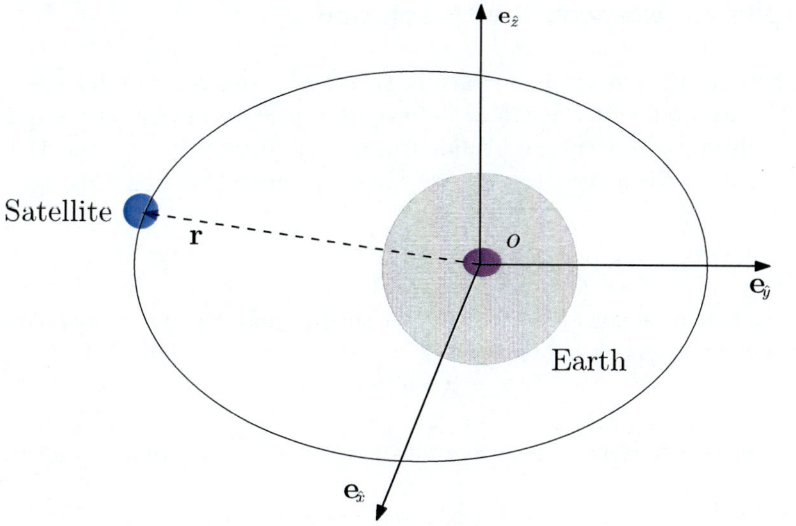

Fig. 2.4 Earth-centered inertial frame

2.1.1 Special Case: Two-Body Problem Consisting of a Large and a Small Body

Next, let us focus on the specific case where one of the bodies is very large as compared to the other one. The motion of a satellite orbiting around the Earth is a typical example. In this context, the larger body B_1 is the Earth and the smaller body B_2 is the satellite. Note that the mass of Earth $m_1 = m_E = 5.972 \times 10^{24}$ kg, whereas the typical mass of a satellite $m_2 = m_S = 5 \times 10^4$ kg or lesser (large satellites such as international space stations are exceptions, but in that case too the typical mass is about 4×10^5 kg). Hence it is quite obvious that $m_E \gg m_S$, i.e., $m_1 \gg m_2$ (in other words, m_2 is negligible as compared to m_1).

Because of the above fact, it can be safely assumed that $m_1 + m_2 \approx m_1$. Moreover, in the case that one orbits around the other, both $\mathbf{r}_1$ and $\mathbf{r}_2$ are finite. Hence, from Eq. (2.18) it is obvious that $\mathbf{r}_1 \simeq 0$, i.e., the center of the coordinate system is nothing but the center of the larger body. Hence, for all practical purposes, Fig. 2.3 can be viewed as in Fig. 2.4, which also includes a typical orbit of the smaller body around the bigger body.

Control force on Earth is obviously negligible. Moreover, perturbation force on Earth is also assumed to be negligible (as compared to those on the satellite) and $(m_E + m_S) \approx m_E$. With these assumptions, Eq. (2.12) for the Earth–satellite pair can be simplified as follows:

$$\ddot{\mathbf{r}} = -\frac{Gm_E}{r^2}\mathbf{e}_r + \frac{\mathbf{F}_{2c}}{m_S} + \frac{\mathbf{F}_{2p}}{m_S} = -\frac{\mu}{r^3}\mathbf{r} + \mathbf{u} + \mathbf{a}_p \tag{2.19}$$

where $\mu \triangleq Gm_E = 398{,}601\,\frac{\text{km}^3}{\text{s}^2}$, $\mathbf{u} \triangleq \frac{\mathbf{F}_{2c}}{m_S}$ (control acceleration) and $\mathbf{a}_p \triangleq \frac{\mathbf{F}_{2p}}{m_S}$ (disturbance acceleration).

Once again, it can be pointed out here that the derived orbital motion of a satellite around the Earth is a restricted two-body problem. The Earth is assumed to be inertially fixed in space. The system model developed in this section also includes the presence of disturbing forces comprising of the gravitational perturbation due to oblateness of Earth (J_2 perturbation), aerodynamic drag, solar radiation pressure, and third body gravitational pull on the satellites.

2.2 The Keplerian Two-body Orbital Solution

In the previous section, the equation of motion for a small body orbiting a large body has been derived. Equation (2.19) is a second-order nonlinear ordinary differential equation. Deriving a closed-form solution considering the control and perturbation effects is a difficult task in general. However, closed-form solution is possible when there is no control and perturbation forces. In that case, Eq. (2.19) can be written as

$$\ddot{\mathbf{r}} = -\frac{\mu}{r^3}\mathbf{r} \tag{2.20}$$

In this context, the specific angular momentum of the smaller body, i.e., the angular momentum of the smaller body per unit mass, is given by

$$\mathbf{h} = \mathbf{r} \times \dot{\mathbf{r}} \tag{2.21}$$

Next, taking the cross product on both sides of Eq. (2.20) with the specific angular momentum $\mathbf{h}$ results in

$$\ddot{\mathbf{r}} \times \mathbf{h} = -\frac{\mu}{r^3}\mathbf{r} \times \mathbf{h} \tag{2.22}$$

However, as $\dot{\mathbf{r}} \times \dot{\mathbf{h}} = 0$ (since $\dot{\mathbf{h}} = 0$), the left-hand side of Eq. (2.22) can be simplified as

$$\ddot{\mathbf{r}} \times \mathbf{h} = \frac{d}{dt}(\dot{\mathbf{r}} \times \mathbf{h}) - \dot{\mathbf{r}} \times \dot{\mathbf{h}} = \frac{d}{dt}(\dot{\mathbf{r}} \times \mathbf{h}) \tag{2.23}$$

Similarly, the right-hand side of Eq. (2.22) can be simplified as

$$\begin{aligned}
\frac{1}{r^3}[\mathbf{r} \times \mathbf{h}] &= \frac{1}{r^3}[\mathbf{r} \times (\mathbf{r} \times \dot{\mathbf{r}})] \\
&= \frac{1}{r^3}[\mathbf{r}(\mathbf{r} \cdot \dot{\mathbf{r}}) - \dot{\mathbf{r}}(\mathbf{r} \cdot \mathbf{r})] \\
&= \frac{1}{r^3}\left[\mathbf{r}(r\dot{r}) - \dot{\mathbf{r}}r^2\right] \\
&= \frac{1}{r^2}[\mathbf{r}\dot{r} - \dot{\mathbf{r}}r)] \\
&= -\frac{d}{dt}\left(\frac{\mathbf{r}}{r}\right)
\end{aligned} \tag{2.24}$$

Substituting Eqs. (2.23) and (2.24) in Eq. (2.22) results in

$$\frac{d}{dt}(\dot{\mathbf{r}} \times \mathbf{h}) = \frac{d}{dt}\left(\mu\frac{\mathbf{r}}{r}\right) \tag{2.25}$$

which can be rewritten as

$$\frac{d}{dt}\left(\dot{\mathbf{r}} \times \mathbf{h} - \mu\frac{\mathbf{r}}{r}\right) = 0 \tag{2.26}$$

Equation (2.26) obviously leads to

$$\dot{\mathbf{r}} \times \mathbf{h} - \mu\frac{\mathbf{r}}{r} = \mathbf{c} \tag{2.27}$$

where $\mathbf{c}$ is a constant vector.

Next, Eq. (2.27) can be rewritten as

$$\frac{1}{\mu}(\dot{\mathbf{r}} \times \mathbf{h}) = \frac{\mathbf{r}}{r} + \mathbf{e} \tag{2.28}$$

where dimensionless vector $\mathbf{e} = \mathbf{c}/\mu$ is called as *eccentricity vector*. To get scalar expression, taking the dot product on both sides with respect to $\mathbf{r}$ results in

$$\frac{1}{\mu}\mathbf{r} \cdot (\dot{\mathbf{r}} \times \mathbf{h}) = \frac{\mathbf{r} \cdot \mathbf{r}}{r} + \mathbf{r} \cdot \mathbf{e} \tag{2.29}$$

Using Eq. (2.21), left-hand side of Eq. (2.29) results in

$$\frac{1}{\mu}\mathbf{r} \cdot (\dot{\mathbf{r}} \times \mathbf{h}) = \frac{1}{\mu}(\mathbf{r} \times \dot{\mathbf{r}}) \cdot \mathbf{h} = \mathbf{h} \cdot \mathbf{h} = h^2 \tag{2.30}$$

Substituting Eq. (2.30) in Eq. (2.29) results in

$$\frac{h^2}{\mu} = r + \mathbf{r} \cdot \mathbf{e} = r + re\ \cos \nu \tag{2.31}$$

where ν is the angle between the eccentricity vector $\mathbf{e}$ and the position vector $\mathbf{r}$, known as the *true anomaly*. And eccentricity e is defined as the magnitude of the eccentricity vector $\mathbf{e}$.

In view of the above discussion, the orbital parameters of the smaller body around the larger body under the influence of the gravity of the larger body are constrained as

$$r = \frac{h^2}{\mu}\left(\frac{1}{1 + e\ \cos \nu}\right) \tag{2.32}$$

Next, for computing the velocity of the smaller body at any given location, taking dot product of Eq. (2.20) with $\dot{\mathbf{r}}$ on both sides gives

$$\ddot{\mathbf{r}} \cdot \dot{r} = -\frac{\mu}{r^3}\mathbf{r} \cdot \dot{\mathbf{r}} \tag{2.33}$$

However,

$$\ddot{\mathbf{r}} \cdot \dot{\mathbf{r}} = \frac{1}{2}\frac{d}{dt}(\dot{\mathbf{r}} \cdot \dot{\mathbf{r}}) = \frac{1}{2}\frac{d}{dt}\mathbf{v} \cdot \mathbf{v} = \frac{d}{dt}\left(\frac{v^2}{2}\right) \tag{2.34}$$

Similarly,

$$\frac{\mu}{r^3}\mathbf{r} \cdot \dot{\mathbf{r}} = \frac{\mu}{r^3}r \cdot \dot{r} = -\frac{d}{dt}\left(\frac{\mu}{r}\right) \tag{2.35}$$

Substituting Eqs. (2.34) and (2.35) in Eq. (2.33) results in

$$\frac{d}{dt}\left(\frac{v^2}{2} - \frac{\mu}{r}\right) = 0 \tag{2.36}$$

i.e.,

$$\frac{v^2}{2} - \frac{\mu}{r} = \varepsilon \tag{2.37}$$

where ε is a constant. Equations (2.32)–(2.37) describes the complete motion of the smaller body orbiting under the influence of the gravity of the larger body.

Equations (2.32)–(2.37) encompass different possibilities of all conic sections, namely ellipse, parabola, and hyperbola depending upon velocity and energy conditions. Even though all possibilities are possible, elliptic orbits are commonly found in most of the satellite applications. Note that when $e = 0$, it results in a circular orbit, i.e., circular orbits are special cases of elliptic orbits. The semimajor axis, a, is half of the longest diameter of an ellipse. Perigee is the point in an orbit at which the orbiting body is closest to the object it orbits. Apogee is the point in the orbit where the orbiting body is furthest from the object it orbits.

A brief summary of commonly used orbital formulas are given below for completeness.

Case 1: Ellipses ($0 \leq e < 1$)

$$a = \frac{r_p + r_a}{2} = \frac{h^2}{\mu}\frac{1}{1 - e^2} \tag{2.38}$$

$$\frac{v^2}{2} - \frac{\mu}{r} = -\frac{\mu}{2a} \tag{2.39}$$

$$T = \frac{2\pi}{\sqrt{\mu}} a^{3/2} \tag{2.40}$$

$$e = \frac{r_a - r_p}{r_a + r_p} \tag{2.41}$$

$$\dot{\nu} = \frac{\sqrt{\mu a \left(1 - e^2\right)}}{r^2} \tag{2.42}$$

$$\ddot{\nu} = \frac{-2\mu e(1 + e\cos\nu)^3 \sin\nu}{a^3\left(1 - e^2\right)^3} \tag{2.43}$$

Case 2: Parabolas ($e = 1$)

$$\frac{v^2}{2} - \frac{\mu}{r} = 0 \tag{2.44}$$

$$v_{esc} = \sqrt{\frac{2\mu}{r}} \tag{2.45}$$

Case 3: Hyperbolas ($e \geq 1$)

$$\frac{v^2}{2} - \frac{\mu}{r} = \frac{\mu}{2a} \tag{2.46}$$

$$a = \frac{h^2}{\mu}\frac{1}{e^2 - 1} \tag{2.47}$$

where r_p and r_a are the radius of the perigee and apogee, respectively, h is angular momentum, μ is the standard gravitational parameter of a celestial body, T is the orbital period, v_{esc} is the escape velocity, and v_o is the speed of a satellite in a circular orbit of radius r. An interested reader can see detail derivations of these in [1].

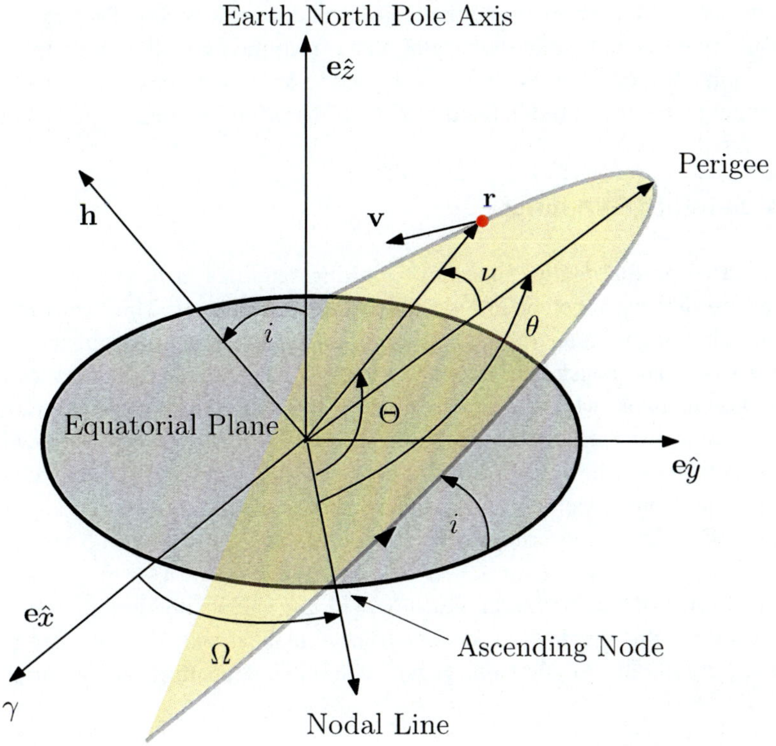

Fig. 2.5 Orbital parameters

A typical elliptic orbit of an Earth-orbiting satellite in three-dimensional space is depicted in Fig. 2.5. Typically Earth-centered inertial (ECI) reference frame is used, which has its origin O at the center of the Earth. ECI frame has $\mathbf{e}_{\hat{x}}$ axis in the direction on the *mean Vernal equinox of epoch J2000.0*, which corresponds to the springtime when Sun crosses the Earth's equator, $\mathbf{e}_{\hat{z}}$ axis is toward North Pole and is perpendicular to the *mean Equator plane (containing the Geocentre) of the epoch J2000.0* and $\mathbf{e}_{\hat{y}}$ axis completes the triad.

To define such an orbit uniquely, one requires the following six orbital elements:

a: semimajor axis,
e: eccentricity,
Ω: right ascension of ascending node,
i: inclination,
θ: argument of perigee, and
ν: true anomaly.

Note that a and e are already defined. Rest of the parameters are defined below. In Fig. 2.5, the intersection of the equatorial plane with the orbital plane is defined as the nodal line. The point on the nodal line at which the satellite crosses the equatorial plane from below is called ascending node while another nodal crossing is called descending node. Also the nodal line vector points outward from the origin through the ascending node. The angle measured along the equatorial plane from the unit vector $\mathbf{e}_{\hat{x}}$ toward the nodal line is called right ascension of ascending node Ω ($0° \leq \Omega \leq 360°$). The specific angular momentum vector h is defined perpendicular to the orbital plane satisfying the right-hand thumb rule to the orbital motion of the satellite. The inclination angle i ($0° \leq i \leq 180°$) is defined as the angle between the equatorial plane and the orbital plane, which is nothing but the angle between the vector $\mathbf{e}_{\hat{z}}$ (i.e., North Pole) and the angular momentum vector. The argument of perigee θ ($0° \leq \theta \leq 360°$) is measured along the orbital plane from the ascending node to the eccentricity vector. The true anomaly ν ($0° \leq \nu \leq 360°$) is defined to know the current angular position of the satellite in the orbital plane, measured in the direction of the orbital motion from the perigee vector.

2.3 Relative Satellite Dynamics

This section is focused on the relative motion of orbiting satellites in formation flying, where two or more satellites are orbiting together in a desired structured manner. This structured manner can be of several configurations such as trailing configuration in the same orbit, cluster configuration in orbits of close proximity and constellation as described in Chapter 1. A leader–follower structure is assumed throughout this book, which is applicable to trailing and cluster configurations. One of the orbiting satellites is known as the *target satellite* or *chief satellite* and other one is known as the *chaser satellite* or *deputy satellite*. In the context of this book, the chief satellite is assumed to be passive or non-maneuvering. The deputy satellite, on the other hand, is assumed to be actively controlled that performs the required maneuver to bring itself into the desired formation with respect to the chief satellite.

In principle, the formation configuration is ensured either by achieving desired true anomaly ν^* for the deputy satellite or by achieving both the desired true anomaly as well as the desired right ascension point Ω^* for the deputy satellite, as shown in Figs. 2.6 and 2.7, depending on the initial condition.

2.3.1 Hill's Reference Frame

In general, the relative distance between the orbiting satellites is so high that considering orbital dynamics in the ECI frame is inconvenient. Instead, the relative dynamics between the orbiting satellites is defined in Hill's reference frame [2] for simplicity, as shown in Fig. 2.8. In fact, this is a non-inertial reference frame centered at the center of gravity of the orbiting chief satellite. This reference frame was first described by G. W. Hill in his original work on the motion of the moon around the Earth [2, 3]. The x-axis of this frame ($\mathbf{e}_x$) is oriented along the radius vector $\mathbf{r}_c$, which is from the center of Earth to the center of this frame. The z-axis ($\mathbf{e}_z$) points in the direction along orbital angular momentum vector

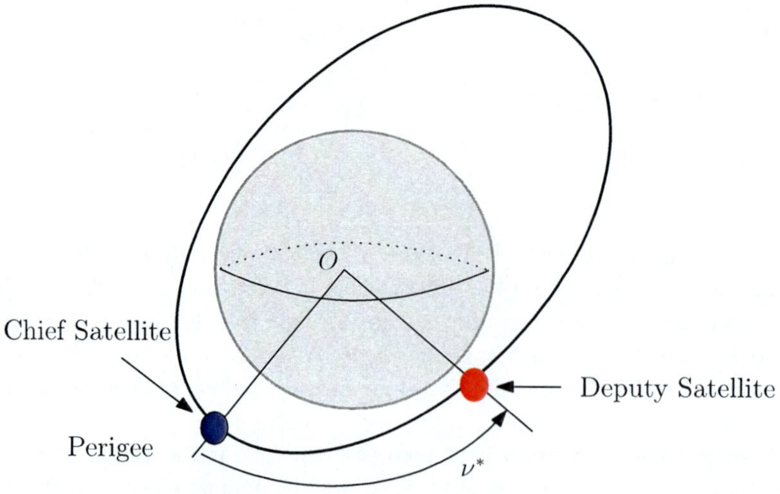

Fig. 2.6 True anomaly shift for formation flying

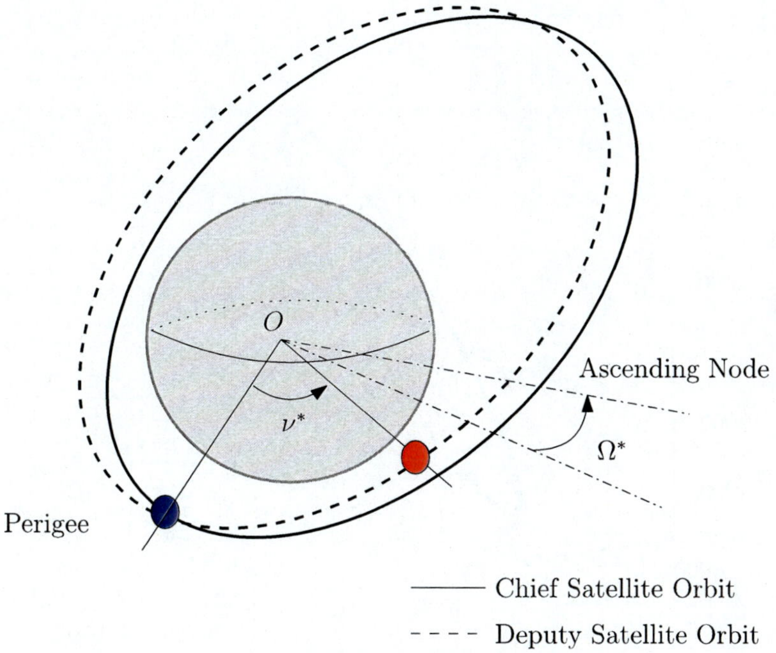

Fig. 2.7 True anomaly and right ascension shift for formation flying

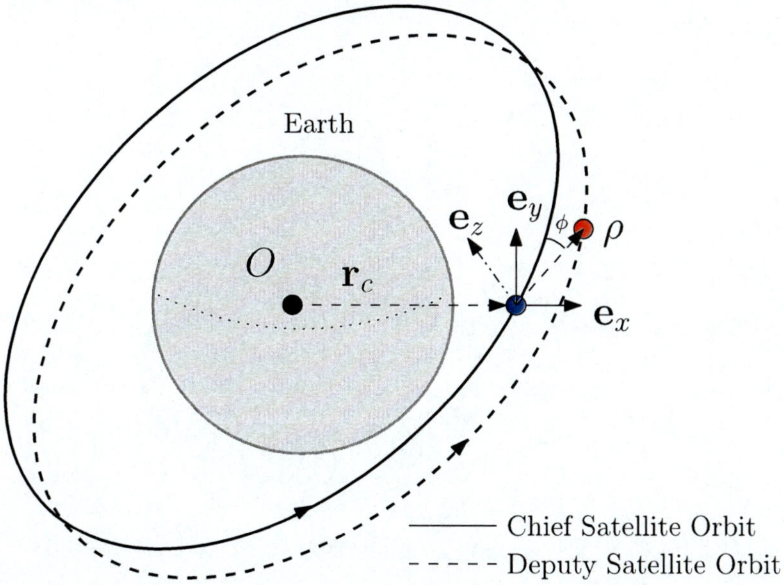

Fig. 2.8 Hill's reference frame for satellite relative motion

($\mathbf{h}$), which is perpendicular to the plane of chief satellite orbit. Finally, the y-axis ($\mathbf{e}_y$) completes the triad [4, 5].

The unit vectors along these axes can be defined as follows:

$$\mathbf{e}_x = \frac{\mathbf{r}_c}{|\mathbf{r}_c|}, \quad \mathbf{e}_z = \frac{\mathbf{h}}{|\mathbf{h}|}, \quad \mathbf{e}_y = \mathbf{e}_z \times \mathbf{e}_x \tag{2.48}$$

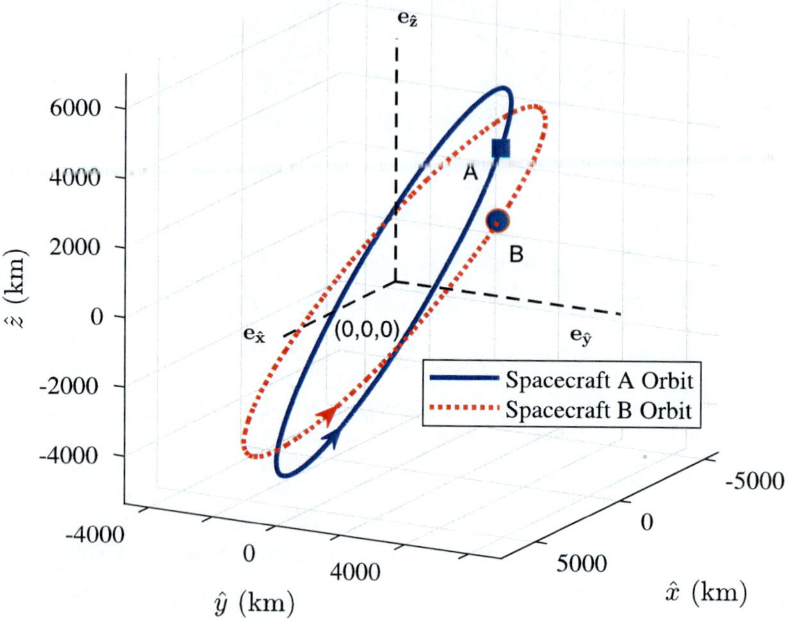

Fig. 2.9 Orbital motion of satellite A and B under the influence of Earth's gravity

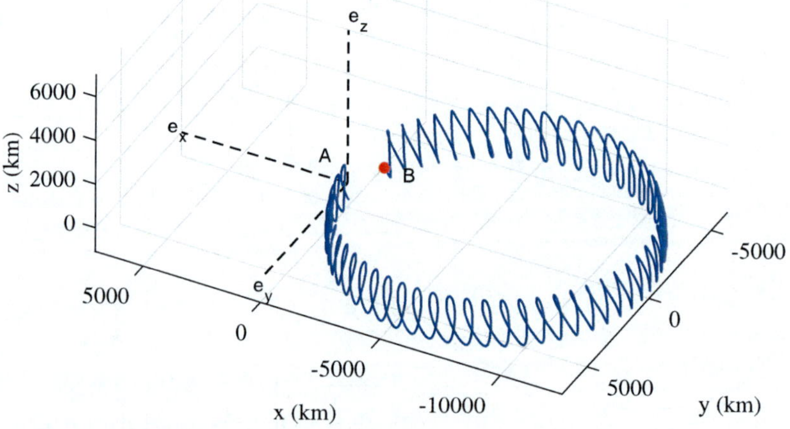

Fig. 2.10 Relative motion of satellite B with respect to A in Hill's Frame

In the formation flying, typically the relative motion of a deputy satellite with respect to the chief satellite is represented in Hill's frame. For better visualization of the resulting trajectories of two such satellites, let us consider two spacecraft A and B having elliptical orbital path around Earth as defined by

$$S_A\text{: } h = 49685\,\text{km/s}^2,\ e = 0.024687,\ i = 59^\circ,\ \Omega = 30^\circ,\ \theta = 20^\circ,\ \nu = 42^\circ$$

$$S_B\colon h = 49985\,\text{km/s}^2,\ e = 0.006954,\ i = 49^\circ,\ \Omega = 30^\circ,\ \theta = 110^\circ,\ \nu = 41^\circ$$

The orbital motion of the two satellites—A and B, in the three-dimensional space of inertial ECI frame is shown in Fig. 2.9. In order to visualize the relative motion of it in Hill's frame, consider spacecraft A as the chief satellite and spacecraft B as the deputy satellite. Figure 2.10 depicts how spacecraft B appears when seen from the origin of Hill's frame while both are in formation flying. For a detailed explanation of the trajectory in Hill's frame, one can refer to Chapter *Relative motion and Rendezvous* of Ref. [1].

2.3.2 Clohessy–Wiltshire Relative Motion in Hill's Frame

In this section, the assumption is that the chief satellite is not under the influence of any control force. Then, as per Eq. (2.19), the equation of motion for the chief satellite in the inertial frame of reference is given as

$$\ddot{\mathbf{r}}_c = -\frac{\mu}{r_c^3}\mathbf{r}_c + \mathbf{a}_{p_c} \tag{2.49}$$

where $\mathbf{r}_c$ is the radius vector of the 'chief satellite' measured from the center of the Earth and $\mathbf{a}_{p_c}$ is the disturbance acceleration on the chief satellite. For circular reference orbit, $|\boldsymbol{r}_c|$ is a constant value, where as for elliptical orbits the instantaneous radius is calculated using Eq. (2.32) as follows:

$$|\mathbf{r}_c| = \frac{a_c\left(1 - e_c^2\right)}{(1 + e_c \cos \nu_c)} \tag{2.50}$$

where a_c is the semimajor axis of chief satellite and e_c is the eccentricity of chief satellite orbit.

Similarly, for the deputy satellite, the equation of motion in the inertial frame is written as

$$\ddot{\mathbf{r}}_d = -\frac{\mu}{r_d^3}\mathbf{r}_d + \mathbf{u} + \mathbf{a}_{p_d} \tag{2.51}$$

where $\mathbf{a}_{p_d}$ is the disturbance acceleration on the deputy satellite. Note that the spatial separation between the chief and the deputy satellite is defined as $\boldsymbol{\rho} = \mathbf{r}_d - \mathbf{r}_c$. Taking double derivative and substituting for $\ddot{\mathbf{r}}_c$ and $\ddot{\mathbf{r}}_d$ from Eqs. (2.49) and (2.51), respectively, yields

$$\ddot{\boldsymbol{\rho}} = -\frac{\mu}{(r_c + \rho)^3}\left(\mathbf{r}_c + \boldsymbol{\rho}\right) + \frac{\mu}{r_c^3}\mathbf{r}_c + \mathbf{u} + \mathbf{a}_{p_r} \tag{2.52}$$

where

$\mathbf{r}_c$ is the radius vector for chief satellite,
$\mathbf{r}_d = (\mathbf{r}_c + \boldsymbol{\rho})$ is the radius vector for deputy satellite, and
$\mathbf{a}_{p_r} = \mathbf{a}_{p_d} - \mathbf{a}_{p_c}$.

Next, the relative acceleration vector $\ddot{\boldsymbol{\rho}}$ is written in the non-inertial Hill's reference frame as follows [1]

$$\ddot{\boldsymbol{\rho}} = \left(\frac{d^2\boldsymbol{\rho}}{dt^2}\right)^{\mathcal{H}} + 2\boldsymbol{\omega}_{\mathcal{I}}^{\mathcal{H}} \times \left(\frac{d\boldsymbol{\rho}}{dt}\right)^{\mathcal{H}} + \left(\frac{d\boldsymbol{\omega}_{\mathcal{I}}^{\mathcal{H}}}{dt}\right) \times \boldsymbol{\rho} + \boldsymbol{\omega}_{\mathcal{I}}^{\mathcal{H}} \times \left(\boldsymbol{\omega}_{\mathcal{I}}^{\mathcal{H}} \times \boldsymbol{\rho}\right) \tag{2.53}$$

where $\boldsymbol{\omega}_{\mathscr{I}}^{\mathscr{H}} = [0\ 0\ \dot{\nu}_c]^T$ denotes the angular velocity of Hill's reference frame relative to inertial reference frame. Substituting Eq. (2.53) in Eq. (2.52) results in

$$\left(\frac{d^2\boldsymbol{\rho}}{dt^2}\right)^{\mathscr{H}} + 2\boldsymbol{\omega}_{\mathscr{I}}^{\mathscr{H}} \times \left(\frac{d\boldsymbol{\rho}}{dt}\right)^{\mathscr{H}} + \left(\frac{d\boldsymbol{\omega}_{\mathscr{I}}^{\mathscr{H}}}{dt}\right) \times \boldsymbol{\rho} + \boldsymbol{\omega}_{\mathscr{I}}^{\mathscr{H}} \times \left(\boldsymbol{\omega}_{\mathscr{I}}^{\mathscr{H}} \times \boldsymbol{\rho}\right)$$
$$-\frac{\mu}{r_c^3}\mathbf{r}_c + \frac{\mu}{r_d^3}\mathbf{r}_d - (\mathbf{u} + \mathbf{a}_{p_r}) = \mathbf{0} \tag{2.54}$$

where $\boldsymbol{\rho} = [x\ y\ z]^T$, $\mathbf{r}_c = [r_c\ 0\ 0]^T$, $\mathbf{r}_d = \boldsymbol{\rho} + \mathbf{r}_c$ and $\mathbf{u} = [a_x\ a_y\ a_z]^T$. Here, x, y, and z are the three component of relative position vector $\boldsymbol{\rho}$ and a_x, a_y, and a_z are applied control accelerations. The term $\mathbf{a}_{p_r} = [a_{p_x}\ a_{p_y}\ a_{p_z}]^T$ includes the external perturbation forces such as gravitational perturbation $\mathbf{a}_{J_{2r}}$, solar radiation pressure, and atmospheric drag in the non-inertial Hill's reference frame. Carrying out the necessary simplifications on Eq. (2.54), the following nonlinear equations are obtained [3].

$$\begin{bmatrix} \ddot{x} \\ \ddot{y} \\ \ddot{z} \end{bmatrix} = \begin{bmatrix} 2\dot{\nu}_c\dot{y} + \ddot{\nu}_c y + \dot{\nu}_c^2 x - \frac{\mu(r_c+x)}{\gamma} + \frac{\mu}{r_c^2} + a_{p_x} \\ -2\dot{\nu}_c\dot{x} - \ddot{\nu}_c x + \dot{\nu}_c^2 y - \frac{\mu y}{\gamma} + a_{p_y} \\ -\frac{\mu z}{\gamma} + a_{p_z} \end{bmatrix} + \begin{bmatrix} 1 & 0 & 0 \\ 0 & 1 & 0 \\ 0 & 0 & 1 \end{bmatrix} \mathbf{u} \tag{2.55}$$

where

$$\gamma \triangleq |r_c + \rho|^3 = \left[(r_c + x)^2 + y^2 + z^2\right]^{\frac{3}{2}} \tag{2.56}$$

The above nonlinear equation of motion is written in control-affine state-space form $\dot{\mathbf{x}} = f(\mathbf{x}) + B\ \mathbf{u}$ as follows:

$$\underbrace{\begin{bmatrix} \dot{x}_1 \\ \dot{x}_2 \\ \dot{x}_3 \\ \dot{x}_4 \\ \dot{x}_5 \\ \dot{x}_6 \end{bmatrix}}_{\dot{\mathbf{x}}} = \underbrace{\begin{bmatrix} x_2 \\ 2\dot{\nu}_c x_4 + \ddot{\nu}_c x_3 + \dot{\nu}_c^2 x_1 - \frac{\mu}{\gamma}(x_1 + r_c) + \frac{\mu}{r_c^2} + \mathrm{a}_{\mathrm{p}_x} \\ x_4 \\ -2\dot{\nu}_c x_2 - \ddot{\nu}_c x_1 + \dot{\nu}_c^2 x_3 - \frac{\mu}{\gamma} x_3 + \mathrm{a}_{\mathrm{p}_y} \\ x_6 \\ -\frac{\mu}{\gamma} x_5 + \mathrm{a}_{\mathrm{p}_z} \end{bmatrix}}_{f(\mathbf{x})} + \underbrace{\begin{bmatrix} 0 & 0 & 0 \\ 1 & 0 & 0 \\ 0 & 0 & 0 \\ 0 & 1 & 0 \\ 0 & 0 & 0 \\ 0 & 0 & 1 \end{bmatrix}}_{B} \mathbf{u} \tag{2.57}$$

where

$$\mathbf{x} \triangleq [x\ \dot{x}\ y\ \dot{y}\ z\ \dot{z}]^T = [x_1\ x_2\ x_3\ x_4\ x_5\ x_6]^T \tag{2.58}$$

2.3.3 Hill's Equation: The Linearized Clohessy–Wiltshire Equation

Hill's Equation is the linearized form of Clohessy–Wiltshire equation of relative motion of two satellites in Hill's frame of Ref. [2]. The linearization of Eq. (2.55) is carried out under the following assumptions:

(i) Circular reference orbit of the chief satellite orbit around the Earth, i.e., $\ddot{\nu}_c = 0$. The angular velocity of the chief satellite $\omega = \dot{\nu}_c = \sqrt{\frac{\mu}{r_c^3}}$ is constant.

(ii) The radial separation between the chief and deputy satellites ($\boldsymbol{\rho}$) is very small as compared to the radius vector $\mathbf{r}_c$ of the chief satellite from the center of Earth ($\boldsymbol{\rho} \ll \mathbf{r}_c$).

(iii) No external perturbations acting on the chief as well as deputy satellites, i.e., $\mathbf{a}_{p_r} = \mathbf{0}$

Using the definition of γ and $\ddot{v}_c = 0$, the Clohessy–Wiltshire equation Eq. (2.55) gets simplified to

$$\begin{aligned} \ddot{x} - 2\omega\dot{y} - \omega^2(r_c + x)\left[1 - \frac{r_c^3}{\left[(r_c + x)^2 + y^2 + z^2\right]^{\frac{3}{2}}}\right] - a_x = 0 \\ \ddot{y} + 2\omega\dot{x} - \omega^2 y\left[1 - \frac{r_c^3}{\left[(r_c + x)^2 + y^2 + z^2\right]^{\frac{3}{2}}}\right] - a_y = 0 \\ \ddot{z} + \omega^2 z\left[\frac{r_c^3}{\left[(r_c + x)^2 + y^2 + z^2\right]^{\frac{3}{2}}}\right] - a_z = 0 \end{aligned} \tag{2.59}$$

Note that Eq. (2.59) is still a nonlinear equation. The nonlinear term is

$$\begin{aligned} \gamma &= \left[(r_c + x)^2 + y^2 + z^2\right]^{\frac{3}{2}} \\ &= r_c^3\left[1 + \frac{2x}{r_c} + \frac{x^2}{r_c^2} + \frac{y^2}{r_c^2} + \frac{z^2}{r_c^2}\right]^{\frac{3}{2}} \end{aligned} \tag{2.60}$$

Using power series expansion, Eq. (2.60) can be written as

$$\gamma = r_c^3\left[1 + \frac{3}{2}\left(\frac{2x}{r_c} + \frac{x^2}{r_c^2} + \frac{y^2}{r_c^2} + \frac{z^2}{r_c^2}\right) + \ldots + HOT\right] \tag{2.61}$$

Neglecting the higher order terms (HOT) and using the second assumption, i.e., $\rho \ll r_c$, it is obvious that $(x^2/r_c^2) \approx 0$, $(y^2/r_c^2) \approx 0$, $(z^2/r_c^2) \approx 0$. With the simplification the nonlinear term γ is written as

$$\gamma = r_c^3\left(\frac{r_c + 3x}{r_c}\right) \tag{2.62}$$

Substituting Eq. (2.62) in Eq. (2.59) and carrying out the further algebraic simplification, the following equation of motion of the relative dynamics is obtained.

$$\begin{aligned} \ddot{x} - 2\omega\dot{y} - \left[\frac{\omega^2(r_c + x)3x}{(r_c + 3x)}\right] - a_x = 0 \\ \ddot{y} + 2\omega\dot{x} - \left[\frac{3\omega^2 yx}{(r_c + 3x)}\right] - a_y = 0 \\ \ddot{z} + \left[\frac{\omega^2 z r_c}{(r_c + 3x)}\right] - a_z = 0 \end{aligned} \tag{2.63}$$

Further, with following approximations

$$
\begin{aligned}
r_c + 3x &\approx r_c \\
r_c + x &\approx r_c \\
\frac{yx}{(r_c + 3x)} &\approx 0
\end{aligned}
\tag{2.64}
$$

the final linearized form of the Clohessy–Wiltshire equation of motion, also known as Hill's equation, can be written as

$$
\begin{aligned}
\ddot{x} &= 2\omega\dot{y} + 3\omega^2 x + a_x \\
\ddot{y} &= -2\omega\dot{x} + a_y \\
\ddot{z} &= -\omega^2 z + a_z
\end{aligned}
\tag{2.65}
$$

Defining the state vector as earlier (see Eq. (2.58)), Hill's equation can be written as

$$
\underbrace{\begin{bmatrix} \dot{x}_1 \\ \dot{x}_2 \\ \dot{x}_3 \\ \dot{x}_4 \\ \dot{x}_5 \\ \dot{x}_6 \end{bmatrix}}_{\dot{\mathbf{x}}} = \underbrace{\begin{bmatrix} 0 & 1 & 0 & 0 & 0 & 0 \\ 3\omega^2 & 0 & 0 & 2\omega & 0 & 0 \\ 0 & 0 & 0 & 1 & 0 & 0 \\ 0 & -2\omega & 0 & 0 & 0 & 0 \\ 0 & 0 & 0 & 0 & 0 & 1 \\ 0 & 0 & 0 & 0 & -\omega^2 & 0 \end{bmatrix}}_{A} \underbrace{\begin{bmatrix} x_1 \\ x_2 \\ x_3 \\ x_4 \\ x_5 \\ x_6 \end{bmatrix}}_{\mathbf{x}} + \underbrace{\begin{bmatrix} 0 & 0 & 0 \\ 1 & 0 & 0 \\ 0 & 0 & 0 \\ 0 & 1 & 0 \\ 0 & 0 & 0 \\ 0 & 0 & 1 \end{bmatrix}}_{B} \mathbf{u}
\tag{2.66}
$$

which takes the standard state-space form [6], i.e., $\dot{\mathbf{x}} = A\mathbf{x} + B\,\mathbf{u}$.

For Hill's equation, under the assumption that there is no control force i.e., $\mathbf{u} = 0$, there exists closed-form solution which is as follows:

$$
\begin{aligned}
x_1 &= \rho \sin(\omega t + \phi) + C_a \\
x_2 &= 2\rho \cos(\omega t + \phi) - \frac{3\omega}{2} C_a t + C_b \\
x_3 &= C_m \rho \sin(\omega t + \phi) + 2C_n \rho \cos(\omega t + \phi) \\
x_4 &= \rho \cos(\omega t + \phi) \\
x_5 &= -2\rho\omega \sin(\omega t + \phi) - \frac{3\omega}{2} C_a \\
x_6 &= C_m \rho\omega \cos(\omega t + \phi) - 2C_n \rho\omega \sin(\omega t + \phi)
\end{aligned}
\tag{2.67}
$$

where C_a and C_b are the center offset of ellipse traced by deputy satellite with respect to chief satellite, ϕ is the angle of deputy satellite position vector with respect to chief satellite velocity vector (refer to Fig. 2.8), C_m and C_n are the slopes of the line formed by the rotation about the minor and major axis, respectively. Note that Eq. (2.67) relates the state parameters of the deputy satellite in Hill's reference frame to the radial separation distance ρ and angular velocity of the chief satellite for any satellite formation flying configuration [2].

2.4 Perturbation Modeling

The two-body problem discussed so far is under the ideal conditions of existence of only those two bodies and their motion is purely based on the influence of gravity of each other. In reality, however, there are many other perturbation forces which affect this motion. Such orbital perturbations are mainly classified into two categories—gravitational and non-gravitational. Gravitational perturbations include gravitational harmonics (which arise because of the bodies being not perfect spheres) and attraction due to third bodies [7]. Non-gravitational perturbations includes atmospheric drag, solar radiation pressure, and tidal effects in case one or both bodies contain large liquid masses (such as water) [7]. The tidal effects arise because of the rotational motion of the interacting bodies about their own axes. Note that only gravitational harmonics due to bodies being not perfect spheres is discussed in detail here as this is the only perturbation considered in the subsequent chapters.

2.4.1 Gravitational Harmonics

The Keplerian two-body motion discussed in Section 2.2 is valid for larger body with point mass or spherical mass distribution. In general, the central body is never a perfect sphere. As a typical example, we can consider our own Earth in which the poles are not single defined points; the equatorial radius is approximately 21 km larger than polar radius [7]; Northern and Southern Hemispheres are not symmetric. The flattening of the poles, small bulging effect at the equator and non-symmetrical hemispheres is known as oblateness of the Earth. Due to this, the gravitational force on the orbiting satellites does not pass through the center of the Earth. It causes a variation in the gravitational pull with respect to the angular position of the orbiting body. This variation is represented using 'gravity potential' G^p, the gradient of which effects the components of the gravity vector. Even though this effect is ideally captured as an infinite series, usually only the dominant second-order term (known as the J_2 harmonic) is kept and the higher order terms are neglected as they are significantly smaller. Details of this perturbation effect are discussed next.

Using Eq. (2.19) and considering no control action on the satellite, the equation of motion is written as follows:

$$\ddot{\mathbf{r}} = -\frac{\mu}{r^3}\mathbf{r} + \mathbf{a}_p \tag{2.68}$$

where $\mathbf{r}$ is measured in the ECI frame, that is, $\mathbf{r} = \hat{x}\ \mathbf{e}_{\hat{x}} + \hat{y}\ \mathbf{e}_{\hat{x}} + \hat{z}\ \mathbf{e}_{\hat{x}}$ and $|r| = \sqrt{\hat{x}^2 + \hat{y}^2 + \hat{z}^2}$, $\mathbf{a}_p = [a_{p_{\hat{x}}}\ a_{p_{\hat{y}}}\ a_{p_{\hat{z}}}]^T$ includes the external perturbation forces in the inertial frame. Using the above definition and rewriting Eq. (2.68) component wise leads to

$$\ddot{\mathbf{r}} = -\frac{\mu}{r^3}\begin{bmatrix}\hat{x}\\ \hat{y}\\ \hat{z}\end{bmatrix} + \begin{bmatrix}a_{p_{\hat{x}}}\\ a_{p_{\hat{y}}}\\ a_{p_{\hat{z}}}\end{bmatrix} \tag{2.69}$$

As mentioned before, the J_2 effect is assumed to be the only external perturbation force considered in this book.[1] With this assumption, Eq. (2.69) can be rewritten as

$$\ddot{\mathbf{r}} = -\frac{\mu}{r^3}\begin{bmatrix}\hat{x}\\ \hat{y}\\ \hat{z}\end{bmatrix} + \begin{bmatrix}a_{J_{2\hat{x}}}\\ a_{J_{2\hat{y}}}\\ a_{J_{2\hat{z}}}\end{bmatrix} \tag{2.70}$$

[1] By measurement of zonal, tesseral, sectorial coefficients [8], it is found that effects of the first term (i.e., second degree) is at least 400 times larger than the rest of the terms for Earth-bound low-orbit satellites. Hence, for such orbits (which is the focus of this book), for all practical purpose, all higher degree terms can be ignored.

where, $a_{J_{2\hat{x}}}, a_{J_{2\hat{y}}}, a_{J_{2\hat{z}}}$ represents the acceleration due to J_2 perturbation effect in the inertial frame. Next, the components of the perturbation force need to be modeled in detail, which is done as follows.

First, the gravitational potential function of arbitrary degree G^p is expressed as[2]

$$G^p = -\frac{\mu}{r}\left\{\sum_{n=2}^{\infty}\left[\left(\frac{R_e}{r}\right)^n J_n P_n(\sin\phi_e) + \sum_{m=1}^{n}\left(\frac{R_e}{r}\right)^n (C_{nm}\cos\varphi + S_{nm}\sin\varphi)\, P_{nm}(\sin\phi_e)\right]\right\} \tag{2.71}$$

where

λ_e is the geographical longitude measure from prime meridian,
ϕ_e is the geocentric latitude of satellite measured from equator,
R_e is the mean Equatorial radius of Earth,
ω_e is the rotation rate of Earth,
t_e is the time since Greenwich meridian lined up with $\mathbf{e}_{\hat{x}}$ axis of ECI,
J_n is the zonal harmonics of degree n and order zero,
$P_n(\sin\phi_e)$ is the Legendre polynomial function of degree n and order 0,
$P_{nm}(\sin\phi_e)$ is the Legendre polynomial function of degree n and order m,
C_{nm} is the tesseral harmonic coefficient for $n \neq m$,
S_{nm} is the sectorial harmonic coefficient for $n = m$, and
$\varphi \triangleq m\lambda + \omega_e t_e$.

Limiting the series in Eq. (2.71) to $n = 2$ results in

$$G^p_{J_2} = -\frac{\mu}{r}\left(\frac{R_e}{r}\right)^2 J_2 P_2(\sin\phi_e) \tag{2.72}$$

where

J_2 is a constant, which is 0.0010826 for Earth,
P_2 is the second-order Legendre polynomial, defined as

$$P_2(\sin\phi_e) = \frac{1}{2}\left(3\sin^2\phi_e - 1\right) \tag{2.73}$$

Substituting Eq. (2.73) in Eq. (2.72) results in

$$G^p_{J_2} = -\frac{\mu R_e^2 J_2}{2r^3}\left(3\sin^2\phi_e - 1\right) \tag{2.74}$$

Using spherical trigonometry identity ($\sin\phi_e = \sin i \sin\Theta$, where $\Theta \triangleq \theta + \nu$ is the argument of latitude of the orbit as shown in Fig. 2.5), Eq. (2.74) can be rewritten as

$$G^p_{J_2} = \frac{\mu R_e^2 J_2}{2r^3}\left(1 - 3\sin^2 i \sin^2\Theta\right) \tag{2.75}$$

Finally, the acceleration components acting on the satellite due to J_2 perturbation is obtained by taking gradient of the potential function Eq. (2.75) with respect to the corresponding coordinates as follows:

[2] This infinite series representing the gravitational potential function is derived from the oblate Earth model, details of which is beyond the scope of this book. An interested reader, however, can see [9] for more details.

$$\mathbf{a}_{J_2} = \begin{bmatrix} \left(\dfrac{\partial G^p_{J_2}}{\partial r}\right) \\ \dfrac{1}{r}\left(\dfrac{\partial G^p_{J_2}}{\partial \Theta}\right) \\ \dfrac{1}{r\sin\Theta}\left(\dfrac{\partial G^p_{J_2}}{\partial i}\right) \end{bmatrix} = -\frac{3\mu R_e^2 J_2}{2r^4}\begin{bmatrix} 1-3\sin^2 i\ \sin^2\Theta \\ \sin^2 i\ \sin 2\Theta \\ \sin 2i\ \sin\Theta \end{bmatrix}$$

$$= -\frac{3\mu R_e^2 J_2}{2}\left[\frac{1}{r^4}\begin{pmatrix} J_{e_{\hat{x}}}(i,\Theta) \\ J_{e_{\hat{y}}}(i,\Theta) \\ J_{e_{\hat{z}}}(i,\Theta) \end{pmatrix}\right] \tag{2.76}$$

where $J_{e_{\hat{x}}} \triangleq 1-3\sin^2 i\ \sin^2\Theta$, $J_{e_{\hat{y}}} \triangleq \sin^2 i\ \sin 2\Theta$ and $J_{e_{\hat{z}}} \triangleq \sin 2i\ \sin\Theta$. For more details on the gradient of a scalar function in spherical coordinates, interested reader can refer [10].

Next, the objective is to derive the perturbation term in Clohessy–Wiltshire relative dynamics in Eq. (2.57) due to J_2 perturbation. As stated earlier in Section 2.3.2, relative J_2 acceleration $\mathbf{a}_{J_{2r}}$ is defined as the difference in J_2 perturbation acceleration of the deputy and chief satellite. Hence, using Eq. (2.76), $\mathbf{a}_{J_{2r}}$ can be written as

$$\begin{aligned}\mathbf{a}_{J_{2r}} &= \mathbf{a}_{J_{2d}} - \mathbf{a}_{J_{2c}} \\ &= -\frac{3\mu R_e^2 J_2}{2}\left\{\frac{1}{r_d^4}\begin{pmatrix} J_{e_{\hat{x}}}(i_d,\Theta_d) \\ J_{e_{\hat{y}}}(i_d,\Theta_d) \\ J_{e_{\hat{z}}}(i_d,\Theta_d) \end{pmatrix} - \frac{1}{r_c^4}\begin{pmatrix} J_{e_{\hat{x}}}(i_c,\Theta_c) \\ J_{e_{\hat{y}}}(i_c,\Theta_c) \\ J_{e_{\hat{z}}}(i_c,\Theta_c) \end{pmatrix}\right\}\end{aligned} \tag{2.77}$$

where (i_d, Θ_d) and (i_c, Θ_c) are the orbital elements of the deputy and chief satellites, respectively.

It is to be noted that Clohessy–Wiltshire dynamics is represented using relative state variables $\mathbf{x}$. In that context, Eq. (2.77) has to be evaluated using $\mathbf{x}$, which is typically done by transforming the relative state variables $\mathbf{x}$ to orbital elements. An invertable transformation matrix $\Sigma(t)$ is introduced in [11] to convert the orbital elements into relative state variables under J_2 perturbation, which is as follows:

$$\delta\boldsymbol{\xi} = [\Sigma(t)]^{-1}\mathbf{x} \tag{2.78}$$

where $\boldsymbol{\xi} \triangleq [a\ \Theta\ i\ e\cos\omega\ e\sin\omega\ \Omega]^T$ is the state vector in the inertial frame and the relative state vector is defined as $\delta\boldsymbol{\xi} \triangleq \boldsymbol{\xi}_d - \boldsymbol{\xi}_c$. Details of the transformation matrix $\Sigma(t)$ is omitted here for brevity, but can be located in Appendix of Ref. [4] by interested readers.

Once the relative state vector in Hill's frame $\mathbf{x}$ is available, the relative state vector in the inertial frame $\delta\boldsymbol{\xi}$ can be computed from Eq. (2.78). Next, by knowing $\boldsymbol{\xi}_c$, $\boldsymbol{\xi}_d = \boldsymbol{\xi}_c + \delta\boldsymbol{\xi}$ is computed. Finally, knowing both $\boldsymbol{\xi}_c$ and $\boldsymbol{\xi}_d$, the relative J_2 perturbation term $\mathbf{a}_{J_{2r}}$ can be computed from Eq. (2.77).

2.4.2 Third Body Gravitational Attractions

The long-term behavior of a satellite orbit is influenced by third body gravitational attractions such as Sun, Moon, and so on. These attractions induce angular rate in the argument of perigee and right ascension of ascending node resulting in the deviation from Keplerian two-body motion. These effects can be used in favor of spacecraft missions such as Frozen orbits for example. More details on it can be found in [8].

2.4.3 Atmospheric Drag

Atmospheric drag is the atmospheric friction force acting opposite to the relative motion of a satellite. It is a non-conservative force and reduces orbital energy continuously. This perturbation is typically considered for the analysis when the orbit perigee height is small. Space missions such as international space station (ISS) and low Earth orbit (LEO) satellites take corrective measures to account for the atmospheric drag forces. Note that some times it can also be used in favor of spacecraft missions involving aero-braking. More details on it can be found in [8].

2.4.4 Solar Radiation Pressure

Solar radiation pressure (SRP) is due to the impinging photon momentum or radiation pressure from the Sun on the space vehicle. The radiation pressure results in forces and torques of small magnitude on the orbiting bodies. In general, it results in the long-term sinusoidal variations of orbit's eccentricity. This effect is generally considered for interplanetary space travel. Solar sailing, an experimental method of spacecraft propulsion, uses radiation pressure from the Sun as a propulsive force. Moreover, in missions with negligible gravity (such as missions involving Halo orbits), this becomes a non-negligible force. More details on the solar radiation pressure can be found in [8]. Note that for LEO orbit missions, the effect of SRP is negligible because of the presence of other dominant perturbation forces.

2.5 Thrust Realization

In the subsequent chapters of this book, several methods will be presented for generating the guidance commands to realize the objective of formation flying. These commands are nothing but the required acceleration components. In general, such commands are generated in some physically meaningful coordinate frames such as Hill's frame, LVLH frame, ECI frame, etc., as described in Section 2.3. In this book, the commands will be generated in Hill's frame. However, these acceleration commands need to be realized by firing the thrusters attached to the satellite body. Without loss of generality, the thrusters can be assumed to be typically in the body frame located at the center of gravity,[3] as depicted in Fig. 2.11, the details of which can be found in [12]. In summary, the required thrust components generated in Hill's frame need to be transformed into the body frame and then realized through the actuation system in the body frame. A variety of thrusters can be used depending on the mission requirement and hardware limitations. An interested reader can refer to [13] for more information about various propulsion options and the associated thrusting mechanisms. A commonly used mechanism for low Earth orbit missions (where satellite formation flying has more relevance) happens to be the reaction control system (RCS)[4] because of its elegance in operation. Hence, a brief summary of thrust realization using the RCS mechanism is described here for completeness.

Note that the RCS system is primarily an on-off actuator that can produce only a fixed magnitude of thrust as an output. Because of this limitation, the required thrust cannot be realized in the sense of zero tracking error. Hence, the common approach followed is to fire these thrusters frequently with appropriate manipulation of the on-off timings so that the required thrust is realized on an average over a small time window. Methods to achieve a suitable thrust in an average sense are discussed

[3] In case the thrusters are mounted in a slightly different frame to account for practical difficulties, another transformation is necessary, but this factor has been ignored for the sake of simplicity of the discussion.

[4] In most of the satellites, the chemical propellant based RCS thrusters (known as 'hot gas thrusters') are used, even though other options (such as 'cold-gas thrusters') are also possible. One can see [14, 15] for more details.

in this section for completeness, along with the associated necessary information, so that it becomes convenient for a practical engineer to get started in that direction. However, the contents of this section are not being used throughout the book and hence an uninterested reader can ignore this section; it will not disturb the flow of the material presented in rest of the book.

As discussed in the Section 2.3.2, the nonlinear equation of motion in Hill's frame can be written as

$$\dot{\mathbf{r}} = \mathbf{v} \tag{2.79}$$

$$\dot{\mathbf{v}} = f(\mathbf{r}, \mathbf{v}) + \mathbf{a} \tag{2.80}$$

where $\mathbf{r}$, $\mathbf{v}$, $\mathbf{a}$ represent the relative position, relative velocity, and required relative acceleration of the deputy satellite in Hill's frame. Note that the components of acceleration '$\mathbf{a}$' in Hill's frame is computed using various advanced guidance (control synthesis) techniques described in this book. To compute the necessary thrust components in the body frame, however, one can notice that $\mathbf{a}$ can be written as

$$\mathbf{a} = \frac{1}{m}\left(A_H(\bar{\mathbf{q}})\mathbf{f}_b\right) \tag{2.81}$$

where $\mathbf{f}_b$ is the net force required in body frame, $A_H(\bar{\mathbf{q}})$ is the transformation matrix from body frame to Hill's frame, $\bar{\mathbf{q}} = [\mathbf{q}\ q_4]^T$ represents the quaternion representation, where $\mathbf{q} = [q_1\ q_2\ q_3]^T$ is the vector part, q_4 is the scalar part of the quaternion, and the norm satisfies $||\bar{\mathbf{q}}|| = 1$. The transformation matrix $A_H(\mathbf{q})$ can be represented as

$$A_H(\bar{\mathbf{q}}) = \begin{bmatrix} q_1^2 - q_2^2 - q_3^2 + q_4^2 & 2\,(q_1q_2 + q_3q_4) & 2\,(q_1q_3 - q_2q_4) \\ 2\,(q_1q_2 - q_3q_4) & -q_1^2 + q_2^2 - q_3^2 + q_4^2 & 2\,(q_2q_3 + q_1q_4) \\ 2\,(q_1q_3 + q_2q_4) & 2\,(q_2q_3 - q_1q_4) & -q_1^2 - q_2^2 + q_3^2 + q_4^2 \end{bmatrix} \tag{2.82}$$

Note that the $A_H(\bar{\mathbf{q}})$ has the orthogonality property and hence the following relation holds good:

$$A_H^{-1}(\bar{\mathbf{q}}) = A_H^T(\bar{\mathbf{q}}) \tag{2.83}$$

For more details on the quaternion, many of its nice properties as well as its utilities in aerospace vehicles in general and spacecraft control in particular, one can refer to [16]. Using Eqs. (2.81)–(2.83), the required net force in the body frame can be evaluated by the following equation:

$$\mathbf{f}_b = m\left[A_H^T(\bar{\mathbf{q}})\right]\mathbf{a} \tag{2.84}$$

where $\mathbf{f}_b$ still contains the desired net thrust components in the body axes in continuous time. However, it needs to be realized using the force generated by the individual thrusters $\mathbf{f}_i$. Note that RCS thrusters typically cannot produce bi-directional thrust and hence a pair of thrusters are necessary along each principal axis. This fact is depicted in Fig. 2.11. For the three principal axes, there are six thrusters $f_1^+, f_1^-, f_2^+, f_2^-, f_3^+, f_3^-$, which are used for providing the necessary thrust in all directions of the body frame $[\,x_b\ y_b\ z_b\,]$. After the evaluation of the required body force using Eq. (2.84) and depending on the sign of the body force $\mathbf{f}_b$ components $\left(f_{x_b},\ f_{y_b},\ f_{z_b}\right)$, the individual thrusters are selected. Without loss of generality, considering the body force component f_{x_b}, the thruster selection in the x_b direction can be written as follows.

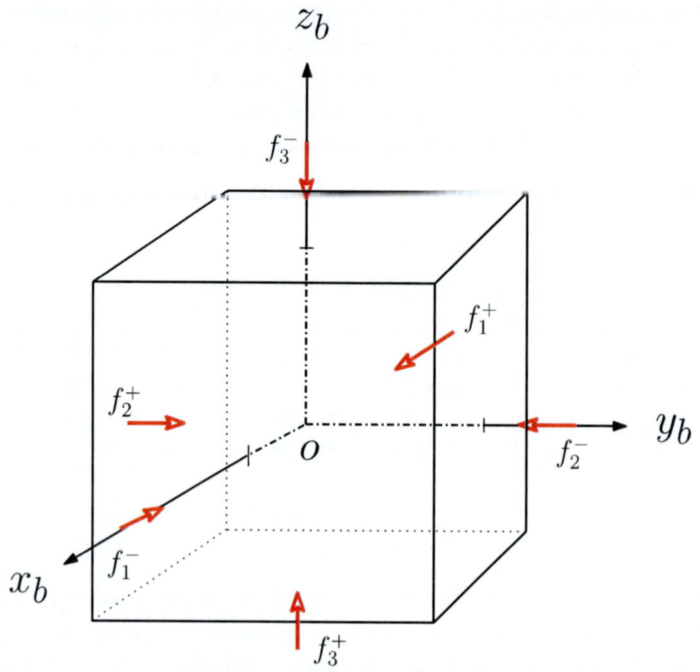

Fig. 2.11 Thruster configuration for a spacecraft

$$
\begin{aligned}
f_1^+ &= f_{x_b},\ f_1^- = 0 \text{ if } f_{x_b} > 0 \\
f_1^- &= f_{x_b},\ f_1^+ = 0 \text{ if } f_{x_b} < 0 \\
f_1^+ &= 0, \quad f_1^- = 0 \text{ if } f_{x_b} = 0
\end{aligned}
\tag{2.85}
$$

The thruster selection logic in y_b and z_b direction operates in a similar fashion.

After the selection is carried out based on the thrust requirements, the selected thruster has to be fired appropriately. As stated before, various techniques discussed in subsequent chapters for generating the guidance commands eventually result in time-varying thruster demand in continuous time. However, the RCS thrusters are non-throttleable engines that can produce only a fixed magnitude of thrust. To address this, there are different pulse modulators are available which translates a continuous demand value to an on/off signal, so that constant magnitude RCS thrusters can meet the overall guidance requirement.

To understand how these pulse modulators operate, some key parameters are defined first. The time duration in which the modulator has a non-zero output is called the thruster on-time T_{on}. The thruster off-time T_{off} is the time duration when the modulator has a zero output. The overall thruster *cycle time* is $T = T_{\text{on}} + T_{\text{off}}$. Next important parameter is the duty cycle DC, which is defined as

$$
DC \triangleq \frac{T_{\text{on}}}{(T_{\text{on}} + T_{\text{off}})}
\tag{2.86}
$$

The duty cycle is a measure of how the modulator responds to an input. It is also used to determine how well the modulator output follows the input. Duty cycle is expressed in percentage; 100% being

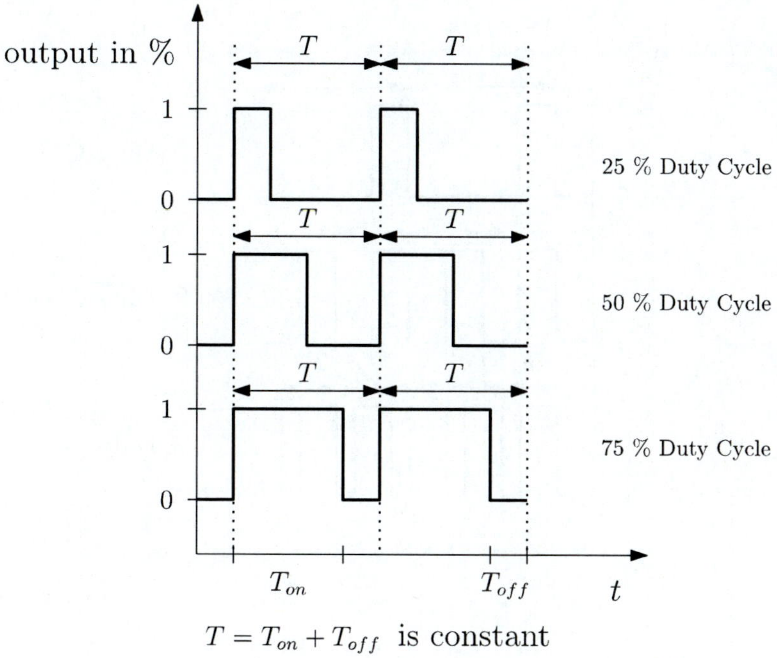

Fig. 2.12 Pulse-width modulator

fully on and 0% being fully off. A low duty cycle obviously corresponds to low thruster on-time as the thruster is off for most of the thruster cycle time.

The key idea of thrust modulation is to select the values of T_{on} and T_{off} to modulate the pulse so that the overall objective is met, i.e., the error between the demanded and realized thrust is minimum. In fact, there are three ways of doing that: (i) pulse-width modulator (PWM), (ii) pulse-frequency modulator (PFM), and (iii) pulse-width pulse-frequency modulator (PWPFM). Only a philosophical explanation is included here for completeness. For rigorous analysis and associated guidelines, one can refer to [17–19].

The PWM modulates the width of its output pulses T_{on} proportionally to the level of thrust commanded at a fixed cycle time T. Figure 2.12 illustrates three different scenarios of duty cycle, namely 25, 50, and 75%, using the PWM concept. Note that the T_{on} value is changed keeping the T as constant to achieve the corresponding duty cycle. The drawback of PWM is when the demand is less, the pulse-width T_{on} decreases keeping the pulse-frequency constant. Since the frequency is constant the RCS thrusters are triggered at a fixed cycle time leading to a decrease in its mechanical efficiency significantly at times of lighter demand. For more details on PWM, interested reader can refer to [20].

In the pulse-frequency modulator (PFM), on the other hand, a continuous signal is converted into a pulse stream, where the pulse frequency $\frac{1}{T}$ is made proportional to the magnitude of thrust commanded. This is done while maintaining a constant pulse width T_{on}. Figure 2.13 illustrates three different scenarios of duty cycle, namely 25, 50, and 75%, using the PFM concept. Note that the value of T is changed, i.e., T_1, T_2, T_3 keeping the value of T_{on} constant to achieve the corresponding duty cycle. Hence, in pulse-frequency modulator, under a heavier demand the pulse frequency increases and under a lighter demand it decreases. On the negative side, however, because the frequency varies, it becomes

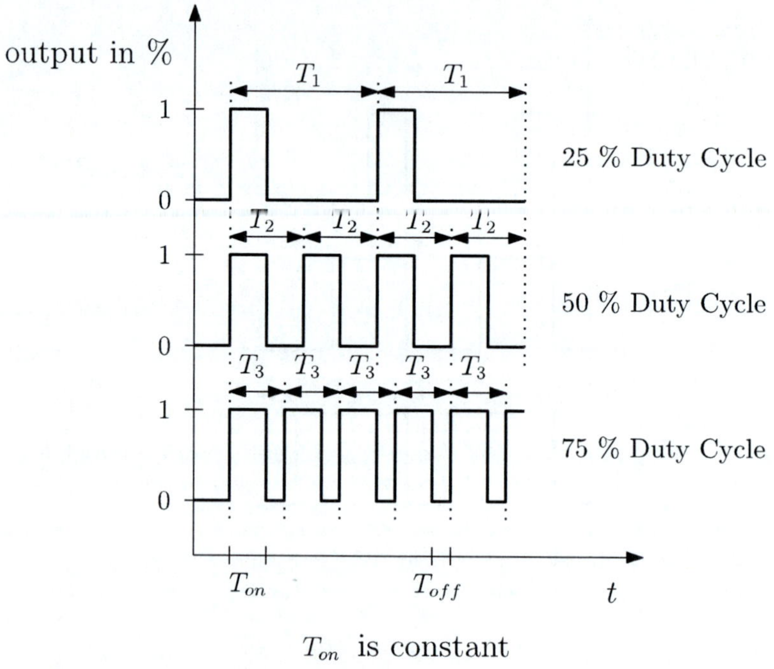

Fig. 2.13 Pulse-frequency modulator

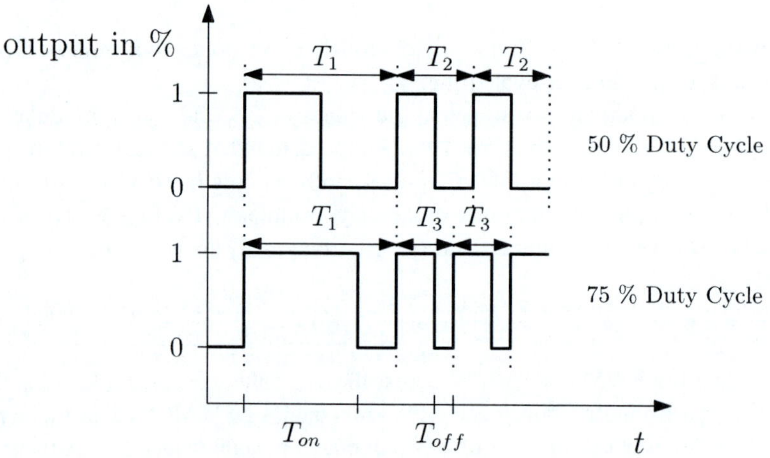

Fig. 2.14 Pulse-width pulse-frequency modulator

sensitive to the noise input. Since noise has high frequency, the RCS also fires at high frequency, which is not a desired feature. For more details on PFM, one can refer to [21].

In order to utilize the merits of PWM and PFM, the pulse-width pulse-frequency modulator (PWPFM) idea has also been proposed in the literature. Here, both T_{on} and T are varied, as shown in Fig. 2.14. Because it performs better than both PWM and PFM, it finds wide usage in practice. Hence, its conceptual operation is discussed next in fair detail for better understanding.

2.5.1 Pulse-Width Pulse Frequency Modulator (PWPFM)

The PWPFM concept explained in the previous section is typically realized as in Fig. 2.15 (see [22] for more details). It contains a first-order lag filter followed by a Schmitt trigger in the forward path and also contains a unity negative feedback loop.

The operational philosophy can be explained as follows:

(i) $U_m(t)$ is initialized to 0. Hence, when $r(t) = 0$ (i.e., there is no input signal), $U_m(t)$ remains at 0.
(ii) With the appearance of $r(t)$, which is assumed to be positive, the filtered error $e_f(t)$ starts increasing. The rate of increase depends on the filter time constant T_m (which serves as a tuning parameter in the design process).
(iii) When $e_f(t)$ increases sufficiently, crossing the Schmitt trigger on-value U_{on}^+, the Schmitt trigger sets in and fixes the output $U_m(t)$ to the prescribed magnitude U_m^+. However, if the value remains below U_{on}^+, the output $U_m(t)$ remains at 0.
(iv) However, this leads to a decrease in the filter input value $e(t)$ due to the negative feedback loop. Depending on the tuning parameters, if there is sufficient decrease until the Schmitt trigger off-value U_{off} is reached, then the output $U_m(t)$ is set to 0.
(v) The same trend can be seen on the negative side as well.

Note that having a dual-limit based switching avoids the chattering issue as well.

To understand the philosophy better, a basic response pattern of a PWPFM is carried out next. In this regard a constant input signal is fed to the modulator, which gives a framework to perform a simple analysis of the modulator behavior for a better intuitive understanding. Also, it gives a good indication on how a spacecraft will behave with a PWPF modulator when implemented in the system. First, the output response of the first-order lag filter in Laplace domain, assuming non-zero initial condition, can be written as

$$e_f(s) = \left[\frac{K_m}{T_m s + 1}\right] e(s) + \left[\frac{T_m}{T_m s + 1}\right] f(0) \tag{2.87}$$

where $e_f(0)$ is the initial condition of the filter, K_m and T_m are the filter gains. The output of the filter feeds the Schmitt trigger, where U_{on} and U_{off} are the Schmitt trigger on/off values and the Schmitt trigger output is $U_m(t)$. The error signal $e(s) = r(s) - U_m(s)$ of the negative feedback loop is the difference between the reference signal $r(s) = C/s$ (assuming $r(t) = C$, a constant) and the modulator

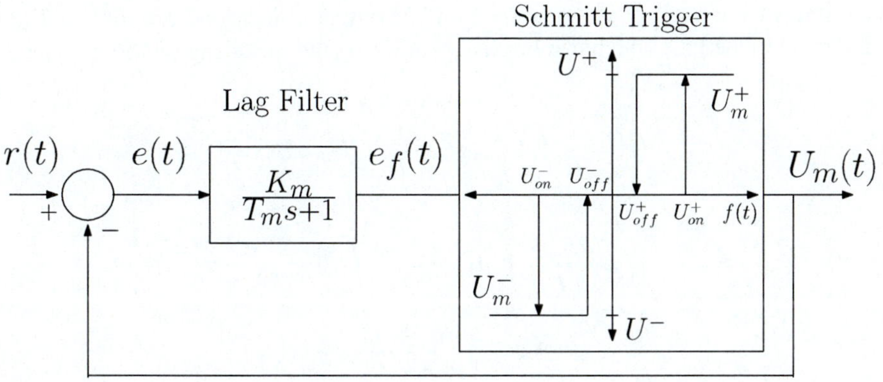

Fig. 2.15 Pulse-width pulse-frequency modulator

output signal $U_m(s) = U_m/s$ (assuming $U_m(t) = U_m$ which is constant as well). Hence Eq. (2.87) can be rewritten as

$$e_f(s) = \left[\frac{K_m}{T_m s + 1}\right] \frac{(C - U_m)}{s} + \left[\frac{T_m}{T_m s + 1}\right] e_f(0) \tag{2.88}$$

Taking inverse Laplace transform and carrying out the necessary algebra, the time domain representation of Eq. (2.88) can be derived as

$$e_f(t) = \left[1 - e^{-\frac{t}{T_m}}\right] K_m(C - U_m) + \left[e^{-\frac{t}{T_m}}\right] e_f(0) \tag{2.89}$$

$$= \left[1 - e^{-\frac{t}{T_m}}\right] \left[K_m(C - U_m) - e_f(0)\right] + e_f(0) \tag{2.90}$$

After the thruster is on, the filter output decreases asymptotically to U_{off}. Hence the initial and final condition of the lag filter for this duration are $e_f(0) = U_{\text{on}}$ and $e_f(T_{\text{on}}) = U_{\text{off}}$, respectively. Using these information, Eq. (2.90) can be written as

$$U_{\text{off}} = \left[1 - e^{-\frac{T_{\text{on}}}{T_m}}\right] [K_m(C - U_m) - U_{\text{on}}] + U_{\text{on}} \tag{2.91}$$

From Eq. (2.91), the thruster on-time T_{on} can be computed as

$$T_{\text{on}} = -T_m \; ln\left[1 - \frac{U_{\text{on}} - U_{\text{off}}}{U_{\text{on}} - K_m(C - U_m)}\right] \tag{2.92}$$

Similarly, the thruster off-time can be computed by substituting $e_f(0) = U_{\text{off}}$ and $e_f(t_{\text{off}}) = U_{\text{on}}$ in Eq. (2.90), which results in

$$U_{\text{on}} = \left[1 - e^{-\frac{T_{\text{off}}}{T_m}}\right] [K_m(C - U_m) - U_{\text{off}}] + U_{\text{off}} \tag{2.93}$$

which in turn leads to

$$T_{\text{off}} = -T_m \; ln\left[1 - \frac{U_{\text{on}} - U_{\text{off}}}{K_m(C - U_m) - U_{\text{off}}}\right] \tag{2.94}$$

For PWPFM, using Eqs. (2.92) and (2.94) in Eq. (2.86), the duty cycle (DC) can be computed. Note that the instantaneous average value of the PWPFM output $\bar{U}$ in a time period $T = T_{\text{on}} + T_{\text{off}}$ is directly proportional to the DC, which can be easily shown by the following relation:

$$\begin{aligned} \bar{U} &= \frac{1}{T} \int_0^{T_{\text{on}}} U_{\text{max}} dt + \frac{1}{T} \int_{T_{\text{on}}}^{T} U_{\text{min}} dt \\ &= \frac{U_{\text{max}}}{T} \int_0^{T_{\text{on}}} dt + \frac{U_{\text{min}}}{T} \int_{T_{\text{on}}}^{T} dt \\ &= \frac{U_{\text{max}}}{T} T_{\text{on}} + \frac{U_{\text{min}}}{T} (T - T_{\text{on}}) \\ &= DC\; U_{\text{max}} + (1 - DC)\, U_{\text{min}} \end{aligned} \tag{2.95}$$

For a RCS thruster, it can be assumed that $U_{\text{max}} = U_m$ and $U_{\text{min}} = 0$, which results in

$$\bar{U} = DC\ U_m \tag{2.96}$$

Hence, for each thruster time cycle, the average thrust realized is a linear fashion of DC. Once the tuning parameters K_m, T_m, U_{on}, U_{off} are selected, the duty cycle is adjusted appropriately at each thruster cycle time by the modulator so that the required thrust is realized. For better conceptual understanding, let us assume that the demanded force is in the form of half of the sine wave from t_0 to t_f that has to be realized by an on-off thruster using the PWPFM philosophy (refer to Fig. 2.16).

According to the Newton's second law of motion, we know that force is equal to the rate of change of linear momentum of an object, i.e.,

$$F(t) = \frac{d}{dt}(mV) \tag{2.97}$$

$$F(t)dt = d(mV) \tag{2.98}$$

where m and V are the mass and velocity of the object, respectively. Taking integral on both sides of Eq. (2.98) from t_0 to t_f results in

$$\Delta(mV) = \int_{t_0}^{t_f} F(t)dt \tag{2.99}$$

The right-hand side of Eq. (2.99) is the area under the curve within the time window $[t_0, t_f]$, shown in blue color in Fig. 2.16. For this input, the PWPFM changes the duty cycle at each thruster cycle time so that the average force realized by the on-off thruster is varied. The area under the realized force using on-off thruster within the time window $[t_0, t_f]$ is shown in red color in Fig. 2.16. One can see the close approximation of the area under the demanded force by the area under the on-off thruster. Hence, from Eq. (2.99) it is clear that the total change in linear momentum $\Delta(mV)$ demanded by the sine wave form of force has been achieved approximately by on-off thruster using PWPFM.

It turns out that mainly PWPFMs are used in spacecraft applications [12] for the following reasons: (i) PWPFM operates close to linear range, i.e., if the required thrust demand goes higher, the realized duty cycle also goes higher, and vice-versa (which is obviously a desirable feature), (ii) A range of parameters are available to tune, which gives a designer the freedom with respect to system-specific considerations (the parameters can be tuned to meet different requirements through different phases

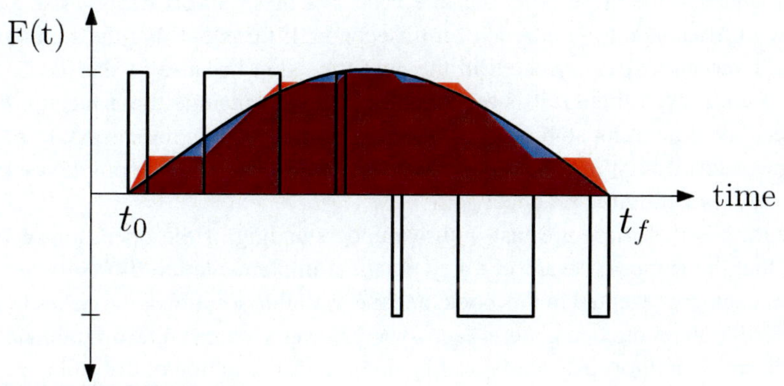

Fig. 2.16 Force approximation by PWPFM

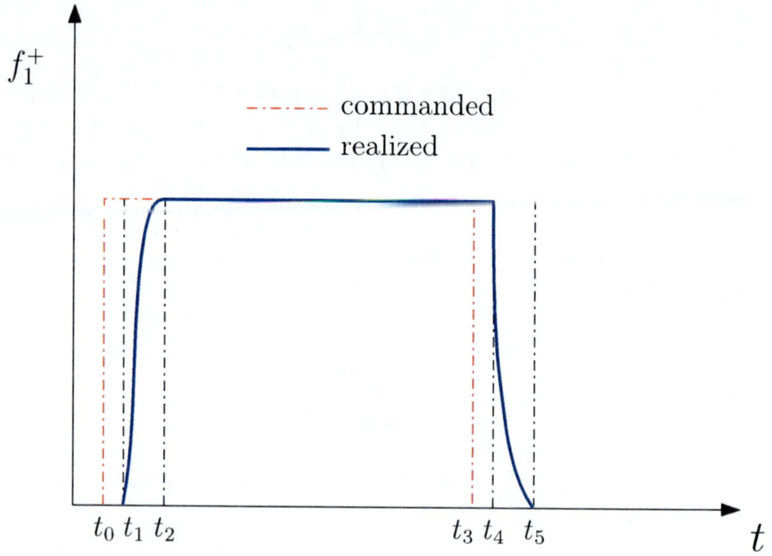

Fig. 2.17 Thrust realization cycle

of operation as well), and (iii) Low fuel consumption and good pointing accuracy, especially when structural vibrations are present. However, the drawback of this modulator is the contribution to the system phase lag which, if not tuned properly, can cause instability to the overall system [17]. A large number of tunable parameters can sometimes lead to difficulty in tuning and/or non-optimal performance. Moreover, the nonlinear characteristics of it make it hard to analyze/compute the stability margins, which are required to characterize the robustness of the closed-loop system [18].

It is obvious to realize that the PWPFM output signal controls the thruster system, thereby resulting in automatic on-off of the RCS system. However, it can also be mentioned here that the inherent delays and dynamics associated with the RCS thruster do not lead to a perfect square form of firing pulses. A typical thrust profile of the realizable RCS thruster f_1^+ firing is shown in Fig. 2.17. First, a force is commanded at time t_0. However, due to the processes required for turning the reaction thruster on, there is a delay, and hence, it actually starts reacting at time t_1. Next, the force from the reaction jet ramps up to the commanded value which is achieved at time t_2. Also, a similar delay is seen when the force command is changed at time t_3, yet the change begins at time t_4. The thrust then ramps down to the new commanded value at t_5. Note that the values of these on-off delays, rise and fall times range from a few milliseconds to hundreds of milliseconds. It turns out that the small errors arising due to these small variations get corrected in the next time step because of the feedback nature of implementation. Hence, even though it is a common practice to include these facts in the extensive six-degree-of-freedom simulation studies, such a high-fidelity model is not necessary to be included in the control (autopilot) design process. This eliminates the mathematical complexity in the design process significantly for negligible practical gain.

A word of caution is that these are just intuitive understanding of the operation of PWM, PFM, and PWPFM so that the reader is aware of a key practical implementation difficulty associated with various guidance schemes presented in this book, and the available solution approaches to overcome it. Because of their utility in practice in general (aerospace, power systems, telecommunication systems, and so on), these have been rigorously studied and guidelines for selection of the tuning parameters are available. An interested reader can see rigorous analysis about PWM and PFM in [20, 21], respectively. Similarly, for more details on the static and dynamic analysis of PWPFM, one can refer to [22].

2.6 Simulation Setup

In order to demonstrate the performance of the guidance techniques discussed in this book, it was aimed to *increase the relative distance* of the deputy satellite to the desired value with respect to the chief satellite. Without loss of generality, this was done in our simulation studies mainly to demonstrate the terminal accuracy better (as it naturally gets amplified).

2.6.1 Formation Flying in a Circular Orbit and with Small Desired Relative Distance

The orbital parameters selected for the chief satellite is circular as tabulated in Table 2.1. The relative conditions of the deputy satellite (refer to Fig. 2.8), i.e., $[\rho, \phi]$ is commanded from the initial value $[1\,\text{km},\ 245°]$ to the final desired value $[5\,\text{km},\ 260°]$. The corresponding relative dynamics states in Hill's frame are tabulated in Table 2.2.

2.6.2 Formation Flying in Elliptic Orbit and with Small Desired Relative Distance

The orbital parameters selected for the chief satellite is assumed to have semimajor axis of 10,000 km and eccentricity of 0.05. The relative position of the deputy satellite $[\rho, \phi]$ is assumed to be guided from the initial value of $[1\,\text{km},\ 245°]$ to the final value of $[5\,\text{km},\ 260°]$. The corresponding relative dynamics states in Hill's frame are tabulated in Table 2.3.

2.6.3 Formation Flying in Circular Orbit and with Large Desired Relative Distance

The orbital parameters selected for the chief satellite is circular as tabulated in Table 2.1. The relative conditions of the deputy satellite, i.e., $[\rho, \phi]$ is commanded from the initial value $[1\,\text{km},\ 245°]$ to the

Table 2.1 Chief satellite orbital parameters

Orbital parameters	Value
Semimajor axis (a)	10,000 km
Eccentricity (e)	0
Orbit inclination (i)	60°
Argument of perigee (ω)	0
Longitude of ascending node (Ω)	0
Initial true anomaly (ν)	0

Table 2.2 Deputy satellite scenario ($\rho_i = 1\,\text{km}$, $\rho_f = 5\,\text{km}$, $e = 0.0$)

Relative states	Initial condition	Final condition
x (km)	-9.06×10^{-1}	-2.32×10^{0}
$\dot{x}$ (km/s)	-2.67×10^{-4}	2.79×10^{-3}
y (km)	-8.45×10^{-1}	8.85×10^{0}
$\dot{y}$ (km/s)	1.14×10^{-3}	2.93×10^{-3}
z (km)	-9.06×10^{-1}	5.37×10^{0}
$\dot{z}$ (km/s)	-2.67×10^{-4}	7.13×10^{-3}

Table 2.3 Deputy satellite scenario ($\rho_i = 1\,\text{km}$, $\rho_f = 5\,\text{km}$, $e = 0.05$)

Relative states	Initial condition	Final condition
x (km)	-9.06×10^{-1}	-3.66×10^{0}
$\dot{x}$ (km/s)	-2.48×10^{-4}	1.70×10^{-3}
y (km)	-8.45×10^{-1}	9.20×10^{0}
$\dot{y}$ (km/s)	1.06×10^{-3}	4.19×10^{-3}
z (km)	-9.06×10^{-1}	5.30×10^{0}
$\dot{z}$ (km/s)	-2.48×10^{-4}	7.18×10^{-3}

Table 2.4 Deputy satellite scenario ($\rho_i = 1\,\text{km}$, $\rho_f = 100\,\text{km}$, $e = 0.0$)

Relative states	Initial condition	Final condition
x (km)	-9.06×10^{-1}	-4.68×10^{1}
$\dot{x}$ (km/s)	-2.67×10^{-4}	5.55×10^{-2}
y (km)	-8.45×10^{-1}	1.76×10^{2}
$\dot{y}$ (km/s)	1.14×10^{-3}	5.77×10^{-2}
z (km)	-9.06×10^{-1}	1.09×10^{2}
$\dot{z}$ (km/s)	-2.67×10^{-4}	1.43×10^{-1}

final desired value $[100\,\text{km},\ 260°]$. The corresponding relative dynamics states in Hill's frame are tabulated in Table 2.4.

2.6.4 Formation Flying in Elliptic Orbit and with Large Desired Relative Distance

The orbital parameters selected for the chief satellite is assumed to have semimajor axis of 10,000 km and eccentricity of 0.05 or 0.2. The relative conditions of the deputy satellite, i.e., $[\rho, \phi]$ is commanded from the initial value $[1\ \text{km},\ 245°]$ to the final desired value $[100\,\text{km},\ 260°]$. The corresponding relative dynamics states in Hill's frame are tabulated in Tables 2.5 and 2.6 respectively.

Table 2.5 Deputy satellite scenario ($\rho_i = 1\,\text{km}$, $\rho_f = 100\,\text{km}$, $e = 0.05$)

Relative states	Initial condition	Final condition
x (km)	-9.06×10^{-1}	-7.39×10^{1}
$\dot{x}$ (km/s)	-2.48×10^{-4}	3.30×10^{-2}
y (km)	-8.45×10^{-1}	1.83×10^{2}
$\dot{y}$ (km/s)	1.06×10^{-3}	8.29×10^{-2}
z (km)	-9.06×10^{-1}	1.08×10^{2}
$\dot{z}$ (km/s)	-2.48×10^{-4}	1.44×10^{-1}

Table 2.6 Deputy satellite scenario ($\rho_i = 1$ km, $\rho_f = 100$ km, $e = 0.2$)

Relative states	Initial condition	Final condition
x (km)	-9.06×10^{-1}	-1.66×10^{2}
$\dot{x}$ (km/s)	-2.03×10^{-4}	-1.19×10^{-2}
y (km)	-8.45×10^{-1}	2.72×10^{2}
$\dot{y}$ (km/s)	8.71×10^{-4}	2.07×10^{-1}
z (km)	-9.06×10^{-1}	1.28×10^{2}
$\dot{z}$ (km/s)	-2.03×10^{-4}	1.54×10^{-1}

2.7 Summary

First section of this chapter has introduced the concept of two-body problem under the influence of the gravitational force of each other. Subsequent sections are concentrated on establishing the concept of relative satellite dynamics, first the nonlinear Clohessy–Wiltshire equation and subsequently the linearized Hill's equation (both of which are valid in Hill's frame of reference). A summary of various perturbation effects such as J_2 perturbation of the primary body, gravitational effects due to other bodies, atmospheric drag, and solar radiation pressure effect are discussed. Note that J_2 perturbation is discussed in detail as it is a significant disturbance force in the near vicinity of the primary body. This chapter gives the necessary basics of the orbital dynamics, which forms the basis of the subsequent chapters of this book.

References

1. Curtis, H.D. 2010. *Orbital mechanics for engineering students*. Elsevier.
2. Hill, G.W. 1878. Researches in Lunar theory. *American Journal of Mathematics*, 1 (1): 5–26, 29–147, 245–260.
3. Clohessy, W.H., and R.S. Wiltshire. 1960. Terminal guidance for satellite rendezvous. *Journal of the Aerospace Sciences* 27 (5): 653–658, 674.
4. Alfriend, K., S.R. Vadali, P. Gurfil, J. How, and L. Brege. 2010. *Spacecraft formation flying: Dynamics, control, and navigation*. Elsevier Astrodynamics Series.
5. Park, H.E., S.Y. Park, and K.H. Choi. 2011. Satellite formation reconfiguration and station keeping using SDRE technique. *Aerospace Science And Technology* 15: 440–452.
6. Friedland, B. 2012. *Control system design: An introduction to state-space methods*. Courier Corporation.
7. Vallado, D.A. 2001. *Fundamentals of astrodynamics and applications*, vol. 12. Springer Science & Business Media.
8. Chobotov, V.A. 2002. *Orbital mechanics*. American Institute of Aeronautics and Astronautics.
9. Irvin, D.J., and D.R. Jacques. 2002. A study of linear versus nonlinear control techniques for the reconfiguration of satellite formations. *Advances in the Astronautical Sciences* 589–608.
10. Matthews, P.C. 1998. Gradient, divergence and curl. In *Vector calculus*, 45–64. Springer.
11. Gim, D.W., and K.T. Alfriend. 2003. State transition matrix of relative motion for the perturbed noncircular reference orbit. *Journal of Guidance, Control, and Dynamics* 26 (6): 956–971.
12. Sidi, M.J. 1997. *Spacecraft dynamics and control: A practical engineering approach*, vol. 7. Cambridge University Press.
13. Sutton, G.P., and O. Biblarz. 2016. *Rocket propulsion elements*. Wiley.
14. Pasand, M., A. Hassani, and M. Ghorbani. 2017. A study of spacecraft reaction thruster configurations for attitude control system. *IEEE Aerospace and Electronic Systems Magazine* 32 (7): 22–39.
15. Adler, S., A. Warshavsky, and A. Peretz. 2005. Low-cost cold-gas reaction control system for the sloshsat FLEVO small satellite. *Journal of Spacecraft and Rockets* 42 (2): 345–351.
16. Kuipers, J.B. 1999. *Quaternions and rotation sequences: A primer with applications to orbits, aerospace, and virtual reality*. Princeton University Press.
17. Wie, B., and C.T. Plescia. 1984. Attitude stabilization of flexible spacecraft during stationkeeping maneuvers. *Journal of Guidance, Control, and Dynamics* 7 (4): 430–436.

18. Buck, N.V. 1996. *Minimum vibration maneuvers using input shaping and pulse-width, pulse-frequency modulated thruster control*. Technical Report, Naval Postgraduate School Monterey, CA.
19. Kazimierczuk, M.K. 2015. *Pulse-width modulated DC-DC power converters*. Wiley.
20. Barr, M. 2001. Pulse width modulation. *Embedded Systems Programming* 14 (10): 103–104.
21. Chen, J. 2007. Determine Buck converter efficiency in PFM mode. *Power Electronics Technology* 33 (9): 28–33.
22. Krovel, T. 2005. Optimal tuning of PWPF modulator for attitude control. Technical Report, Master Thesis, Norwegian University of Science and Technology.

Check for updates

Infinite-Time LQR and SDRE for Satellite Formation Flying

3

In this chapter, we demonstrate the applicability of the standard linear quadratic regulator (LQR) and state-dependent Riccati Equation (SDRE) based on linear and nonlinear optimal controllers to achieve the objective of satellite formation flying. Infinite-time formulations of LQR and SDRE offer the advantage of simplicity naturally and hence they are widely preferred in various applications. The utility of these methods for satellite formation flying is demonstrated in their respective validity regions. Both methods are effective when the chief satellite is in the circular orbit and the relative separation distance is small. On the other hand, the SDRE-based formulation is found to be effective even for eccentric orbits, but only with small eccentricity.

3.1 Linear Quadratic Regulator (LQR): Generic Theory

In the LQR theory [1, 2], a quadratic performance index of the following form is minimized:

$$J = \frac{1}{2}\mathbf{x}^T(t_f)S_f\,\mathbf{x}(t_f) + \frac{1}{2}\int_0^{t_f}\left(\mathbf{x}^T Q\mathbf{x} + \mathbf{u}^T R\mathbf{u}\right)dt \tag{3.1}$$

subject to a linear approximation of the plant model, which can be written as

$$\dot{\mathbf{x}} = A\mathbf{x} + B\mathbf{u} \tag{3.2}$$

where $A \in \Re^{n\times n}$ and $B \in \Re^{n\times m}$ are system matrices and t_f is the final time. Q and R are positive semi-definite and positive definite state and control weighting matrices, respectively, which needs to be selected by a control designer judiciously. S_f is also a positive semi-definite weighting matrix on the terminal states, which needs to be selected by the designer. The theory demands that two necessary conditions need to be satisfied, which are (i) the pair $\{A, B\}$ is controllable and (ii) the pair $\{\sqrt{Q}, A\}$

Electronic supplementary material The online version of this chapter (https://doi.org/10.1007/978-981-15-9631-5_3) contains supplementary material, which is available to authorized users.

S. Mathavaraj and R. Padhi, *Satellite Formation Flying*,
https://doi.org/10.1007/978-981-15-9631-5_3

is detectable. Note that in general A, B, Q, and R can be time-varying, and the initial condition $\mathbf{x}(0)$ is assumed to be available.

The minimization of the cost function in Eq. (3.1) is achieved with the following solution of the optimal control expression [1]

$$\mathbf{u}(t) = -K(t)\mathbf{x} = -R^{-1}B^T P(t)\mathbf{x} \tag{3.3}$$

where $P(t)$ satisfies

$$PA + A^T P - PBR^{-1}B^T P + Q = -\dot{P} \tag{3.4}$$

This equation is known as the differential Ricatti equation (DRE). Because it is a matrix differential equation, it also needs a boundary condition, which is computed as

$$P(t_f) = S_f\mathbf{x}(t_f) \tag{3.5}$$

Since the boundary condition is given at the final time, the usual recommendation is to numerically integrate Eq. (3.4) backward offline, starting from t_f using the boundary condition in Eq. (3.5), until $t = 0$ and store the results in the memory. Next, in online usage, the solution $P(t)$ that is available in the memory is used in Eq. (3.3) to compute the optimal control.

For time-invariant systems, A and B are constant matrices. For such systems, if one also selects constant weight matrices Q and R, and, as a special case allows $t_f \to \infty$, then the terminal penalty in Eq. (3.1) is no longer needed and the optimal control expression gets further simplified to

$$\mathbf{u} = -K\mathbf{x} = -R^{-1}B^T P\mathbf{x} \tag{3.6}$$

where P satisfies

$$PA + A^T P - PBR^{-1}B^T P + Q = 0 \tag{3.7}$$

This equation is known as the algebraic Ricatti equation (ARE). Because it is no longer a dynamic equation, it does not need a boundary condition. However, even though it is an algebraic equation, the ARE is a *nonlinear matrix equation* and hence admits multiple solutions for P. One needs to solve for a positive definite solution for P. This is because the positive definite condition on Ricatti matrix P in turn guarantees that the closed-loop system becomes asymptotically stable. Given the condition that the pair $\{A, B\}$ is controllable and the pair $\{\sqrt{Q}, A\}$ is detectable, the solution to ARE admits one and only one positive definite solution.

More details about the LQR theory can be found in numerous textbooks (see, for example, [1, 2]).

3.2 Satellite Formation Flying Control Using LQR

As derived in Section 2.3.3, the linearized equations of motion for the relative dynamics between the deputy and chief satellites can be written as

$$\dot{\mathbf{x}} = A\mathbf{x} + B\mathbf{u} \tag{3.8}$$

where the system matrices A and B are reproduced below for convenience

$$A = \begin{bmatrix} 0 & 1 & 0 & 0 & 0 & 0 \\ 3\omega^2 & 0 & 0 & 2\omega & 0 & 0 \\ 0 & 0 & 0 & 1 & 0 & 0 \\ 0 & -2\omega & 0 & 0 & 0 & 0 \\ 0 & 0 & 0 & 0 & 0 & 1 \\ 0 & 0 & 0 & 0 & -\omega^2 & 0 \end{bmatrix}, \quad B = \begin{bmatrix} 0 & 0 & 0 \\ 1 & 0 & 0 \\ 0 & 0 & 0 \\ 0 & 1 & 0 \\ 0 & 0 & 0 \\ 0 & 0 & 1 \end{bmatrix}$$

Satellite formation flying (SFF) is basically a tracking problem, where the deputy satellite states $\mathbf{x}(t)$ has to track the desired states $\mathbf{x}_d(t)$ with respect to the chief satellite. Note that $\mathbf{x}_d(t)$ needs to be generated in such a way that the deputy satellite keeps flying in a desired 'neighboring orbit', conforming to the orbital parameters of that orbit. The necessary discussion on the topic can be found in Section 2.3.2.

Since the LQR essentially relies on the small perturbation theory and in principle, attempts to nullify the perturbation, this method is a natural choice for SFF, provided (i) the desired orbit of the deputy satellite is in the close vicinity of the chief satellite and (ii) the chief satellite is in a circular orbit.

To proceed on this philosophy, first the '*error state vector*' is defined as $\tilde{\mathbf{x}} \triangleq \mathbf{x} - \mathbf{x}_d$. With this definition, the error state dynamics can be written as

$$\begin{aligned} \dot{\tilde{\mathbf{x}}} &= \dot{\mathbf{x}} - \dot{\mathbf{x}}_d \\ &= (A\mathbf{x} + B\mathbf{u}) - (A\mathbf{x}_d) \\ &= A\tilde{\mathbf{x}} + B\mathbf{u} \end{aligned} \tag{3.9}$$

Equation (3.9) relies on two facts: (i) on the 'desired orbit', the deputy satellite needs no control action and (ii) the system matrix A remains the same for both the perturbed orbit as well as on the desired orbit (as it only depends on the orbital parameters of the chief satellite).

One can notice that Eq. (3.9) represents a linear system dynamics, where the control $\mathbf{u}$ should be designed in such a way that $\tilde{\mathbf{x}} \to 0$. Taking the help of the LQR theory and assuming $t_f \to \infty$, an associated quadratic cost function to be minimized is selected as

$$J = \frac{1}{2} \int_0^\infty \left(\tilde{\mathbf{x}}^T Q \tilde{\mathbf{x}} + \mathbf{u}^T R \mathbf{u} \right) dt \tag{3.10}$$

where Q and R are selected as positive semi-definite and positive definite matrices, respectively. The exact selection of the numerical values used to generate the results is provided in Section 3.5.

Using Eqs. (3.9) and (3.10) and following the theory discussed in Section 3.1, the optimal control is evaluated as

$$\mathbf{u} = -K\,\tilde{\mathbf{x}} \tag{3.11}$$

where the gain matrix K is computed from Eqs. (3.6) and (3.7).

To the best of the knowledge of the authors, this approach was first proposed in [3] and more details can be found there.

3.3 Infinite-Time SDRE: Generic Theory

In Sects. 3.1 and 3.2, the generic LQR theory is presented, followed by its application to the satellite formation flying. However, the theory strongly relies on the linear system dynamics and hence is valid only for close formations in circular orbits. Unfortunately, the approach fails if either of these two conditions are violated. To address this issue to a limited extent, two formulations based on the infinite-time SDRE guidance schemes are presented in this section, which enlarges the application domain. The two SDRE formulations described here differ in the state-dependent coefficient (SDC) forms that they rely on.

Note that the SDRE technique is an intuitive extension of the LQR design under certain assumptions. The key idea here is to describe the system dynamics in linear-looking SDC form. Next, the control solution is derived by repeatedly solving the corresponding Riccati equation online at every grid point of time utilizing this SDC form of system dynamics.

In the SDRE approach, the objective is to regulate a control-affine nonlinear system having the following system dynamics:

$$\dot{\mathbf{x}} = f\left(\mathbf{x}\right) + B\left(\mathbf{x}\right)\mathbf{u} \tag{3.12}$$

about the zero equilibrium point. This is done by minimizing the following quadratic performance index:

$$J = \frac{1}{2}\int_{t_0}^{\infty}\left(\mathbf{x}^{\mathbf{T}}Q\left(\mathbf{x}\right)\mathbf{x} + \mathbf{u}^{\mathbf{T}}R\left(\mathbf{x}\right)\mathbf{u}\right)dt \tag{3.13}$$

The key philosophy in the infinite-time SDRE (I-SDRE) technique is to write the nonlinear function $f\left(\mathbf{x}\right)$ in the state-dependent coefficient (SDC) form $A\left(\mathbf{x}\right)\mathbf{x}$, after which the system dynamics Eq. (3.12) can be written as

$$\dot{\mathbf{x}} = A\left(\mathbf{x}\right)\mathbf{x} + B\left(\mathbf{x}\right)\mathbf{u} \tag{3.14}$$

On close observation, Eqs. (3.13) and (3.14) 'appear to be' in the LQR form, even though it is truly not. With this observation, the solution from the LQR theory is directly applied to propose the following solution for the control variable

$$\mathbf{u} = -\left[R^{-1}\left(\mathbf{x}\right)B^{T}\left(\mathbf{x}\right)P\left(\mathbf{x}\right)\right]\mathbf{x} = -K\left(\mathbf{x}\right)\mathbf{x} \tag{3.15}$$

where the Riccati matrix $P(\mathbf{x})$ is repeatedly computed from the following algebraic Riccati equation:

$$P\left(\mathbf{x}\right)A\left(\mathbf{x}\right) + A^{T}\left(\mathbf{x}\right)P\left(\mathbf{x}\right) - P\left(\mathbf{x}\right)B\left(\mathbf{x}\right)R^{-1}\left(\mathbf{x}\right)B^{T}\left(\mathbf{x}\right)P\left(\mathbf{x}\right) + Q\left(\mathbf{x}\right) = 0 \tag{3.16}$$

Note that the following assumptions need to be valid for the success of the SDRE approach:

(i) $f(0) = 0$
(ii) $B(\mathbf{x}) \neq 0 \quad \forall \mathbf{x} \in \Omega$ (i.e., the domain of interest),
(iii) $Q(\mathbf{x}) \geq 0$ (positive semi-definite), $R(\mathbf{x}) > 0$ (positive definite) $\forall \mathbf{x} \in \Omega$,
(iv) $\{A(\mathbf{x}), B(\mathbf{x})\}$ is point-wise stabilizable $\forall \mathbf{x} \in \Omega$,[1]
(v) $f(\mathbf{x}), B(\mathbf{x}), Q(\mathbf{x}), R(\mathbf{x}) \in C^k \quad (k \geq 1)$.

where $f(\mathbf{x}) \in C^k$ implies that kth-order partial derivative components of $f(\mathbf{x})$ are continuous. Upon close examination, it can be observed that the above procedure does not involve any linearization process. As soon as numerical values of the state vector $\mathbf{x}$ at a grid point of time is inserted in matrices of Eqs. (3.13) and (3.14) and the required control is computed from Eqs. (3.15) and (3.16). Note that

[1] *Stabilizability* is weaker condition of *controllability*, where the uncontrolled modes are stable [4].

Eq. (3.16), which is a nonlinear matrix equation, has to be solved at each grid point of time as the matrix values keep on changing depending on the current value of $\mathbf{x}$. If possible, it is solved in closed form by long hand algebra, but that is mostly possible only for trivial problems. In almost all practical problems, it is solved using numerical algorithms [5]. There are certain nice properties of the SDRE technique [6] as well, which makes the theory fairly rigorous. Those can be summarized as follows:

(i) For scalar problems, the resulting SDRE nonlinear controller satisfies all the necessary conditions of optimality and hence results in an optimal controller. However this is not true for the vector case.
(ii) Under the assumptions listed above, the SDRE approach always produces a closed-loop system that is locally asymptotically stable.
(iii) It results in a suboptimal controller. This sub-optimality is attributed to the fact that the costate equation is satisfied only asymptotically. However, the optimal control equation is always satisfied. Note that the state, costate, and optimal control equations are the three necessary conditions for optimality [1].

It turns out that the non-uniqueness of the SDC parameterization of the system dynamics poses a major challenge in successful implementation of the SDRE technique. It may also restrict the validity domain of the resulting controller. Nevertheless, it has been successfully applied in a number of challenging practical problems. For a comprehensive review of the SDRE technique, one can refer to [7].

3.4 SDC Formulation for Satellite Formation Flying

As mentioned in Section 3.3, the SDRE technique requires that $f(\mathbf{x})$ of the system dynamics is first written in the state-dependent coefficient (SDC) form $A(\mathbf{x})\,\mathbf{x}$. The SDC form, however, is not unique. Moreover, even though all SDC forms result in stabilizing suboptimal controllers, the performance of a particular SDRE controller strongly depends on chosen SDC form. Unfortunately, it remains as an 'art' of the designer to propose a good SDC form for a given problem that leads to good performance. Two SDC formulations for the nonlinear Clohessy–Wiltshire Eq. (2.55), which represents the nonlinear relative dynamics between two satellites which are used to generate the simulation results, are discussed in this section with all necessary details for completeness.

3.4.1 SDC Formulation—1

The first approach, which leads to the state-dependent coefficient—1 (SDC_1) formulation, can be found in [8]. Here the Clohessy–Wiltshire equation (2.55) is simplified using the assumption that the chief satellite is in the circular orbit, i.e., $\ddot{\nu}_c = 0$, which results in the following equations:

$$\ddot{x} - 2\omega\dot{y} - \omega^2(r_c + x)\left[1 - \frac{r_c^3}{\left[(r_c + x)^2 + y^2 + z^2\right]^{\frac{3}{2}}}\right] - a_x = 0 \tag{3.17}$$

$$\ddot{y} + 2\omega\dot{x} - \omega^2 y\left[1 - \frac{r_c^3}{\left[(r_c + x)^2 + y^2 + z^2\right]^{\frac{3}{2}}}\right] - a_y = 0 \tag{3.18}$$

$$\ddot{z} + \omega^2 z\left[\frac{r_c^3}{\left[(r_c + x)^2 + y^2 + z^2\right]^{\frac{3}{2}}}\right] - a_z = 0 \tag{3.19}$$

For simplicity, Eqs. (3.17)–(3.19) can be rewritten as

$$\ddot{x} = 2\omega\dot{y} + \omega^2\sigma_x x + a_x \tag{3.20}$$

$$\ddot{y} = -2\omega\dot{x} + \omega^2\sigma_y y + a_y \tag{3.21}$$

$$\ddot{z} = -\omega^2\sigma_z z + a_z \tag{3.22}$$

where $\sigma_z \triangleq r_c^3/\left[(r_c + x)^2 + y^2 + z^2\right]^{\frac{3}{2}}$, $\sigma_y \triangleq 1 - \sigma_z$ and $\sigma_x \triangleq \left(\frac{r_c}{x} + 1\right)\sigma_y$. Equations (3.20)–(3.22) can be written in the state-space form as

$$\begin{bmatrix} \dot{x}_1 \\ \dot{x}_2 \\ \dot{x}_3 \\ \dot{x}_4 \\ \dot{x}_5 \\ \dot{x}_6 \end{bmatrix} = \underbrace{\begin{bmatrix} 0 & 1 & 0 & 0 & 0 & 0 \\ \omega^2\sigma_x & 0 & 0 & 2\omega & 0 & 0 \\ 0 & 0 & 0 & 1 & 0 & 0 \\ 0 & -2\omega & \omega^2\sigma_y & 0 & 0 & 0 \\ 0 & 0 & 0 & 0 & 0 & 1 \\ 0 & 0 & 0 & 0 & -\omega^2\sigma_z & 0 \end{bmatrix}}_{A} \underbrace{\begin{bmatrix} x_1 \\ x_2 \\ x_3 \\ x_4 \\ x_5 \\ x_6 \end{bmatrix}}_{x} + \underbrace{\begin{bmatrix} 0\,0\,0 \\ 1\,0\,0 \\ 0\,0\,0 \\ 0\,1\,0 \\ 0\,0\,0 \\ 0\,0\,1 \end{bmatrix}}_{B} \boldsymbol{u} \tag{3.23}$$

where $\mathbf{x} \triangleq [x\ \dot{x}\ y\ \dot{y}\ z\ \dot{z}]^T = [x_1\ x_2\ x_3\ x_4\ x_5\ x_6]^T$.

It is to be noted that SDC_1 formulation [8] approximates the nonlinear relative equation motion under the assumption that chief satellite orbit is circular, i.e., $\ddot{\nu}_c = 0$. Therefore SDC representation in Eq. (3.23) only caters to the circular reference orbit scenario of satellite formation flying. When this assumption is not true, it leads to erroneous results. This necessitates the exploration of alternate SDC formulations, one of which is discussed next.

3.4.2 SDC Formulation—2

The state-dependent coefficient formulation—2 (SDC_2) is available in [9], which turns out to be a better representation because it does not require the chief satellite to be in a circular orbit. The details of this formulation are presented now.

First, the nonlinear term in Eq. (2.55) is simplified as

$$\begin{aligned} \frac{\mu}{\gamma} r_c - \frac{\mu}{r_c^2} &= \mu\left[\frac{r_c}{\left[(r_c + x)^2 + y^2 + z^2\right]^{3/2}} - \frac{1}{r_c^2}\right] \\ &= \mu\left[\frac{r_c}{\left[r_c^2 + 2r_c x + x^2 + y^2 + z^2\right]^{3/2}} - \frac{1}{r_c^2}\right] \end{aligned} \tag{3.24}$$

Factorizing the term r_c^2 from the denominator gives

$$\frac{\mu}{\gamma}r_c - \frac{\mu}{r_c^2} = \frac{\mu}{r_c^2}\left[\frac{1}{\left[1+2\frac{x}{r_c}+\frac{x^2}{r_c^2}+\frac{y^2}{r_c^2}+\frac{z^2}{r_c^2}\right]^{\frac{3}{2}}} - 1\right]$$
$$= \frac{\mu}{r_c^2}\left[\left[1-\left(-2\frac{x}{r_c}-\frac{(x^2+y^2+z^2)}{r_c^2}\right)\right]^{-\frac{3}{2}} - 1\right]$$
$$= \frac{\mu}{r_c^2}\left[[1-\xi]^{-\frac{3}{2}} - 1\right] \tag{3.25}$$

where,

$$\xi = -2\frac{x}{r_c} - \frac{(x^2+y^2+z^2)}{r_c^2}$$
$$= \left(-\frac{2}{r_c}-\frac{x}{r_c^2}\right)x + \left(-\frac{y}{r_c^2}\right)y + \left(-\frac{z}{r_c^2}\right)z \tag{3.26}$$

Next, by employing the standard negative binomial expansion[2] [10], the term $[1-\xi]^{-\frac{3}{2}}$ in Eq. (3.25) can be written as

$$(1-\xi)^{-\frac{3}{2}} = 1 + \frac{3}{2}\xi + \frac{3}{2}\frac{\left(\frac{3}{2}+1\right)}{2!}\xi^2 + \frac{3}{2}\frac{\left(\frac{3}{2}+1\right)\left(\frac{3}{2}+2\right)}{3!}\xi^3 + \cdots \tag{3.28}$$

Using this series expression, the left-hand side term in Eq. (3.25) can be expressed as

$$\frac{\mu}{\gamma}r_c - \frac{\mu}{r_c^2} = \frac{\mu}{r_c^2}\left[1+\frac{3}{2}\psi\xi - 1\right] = \frac{3\mu}{2r_c^2}\psi\xi \tag{3.29}$$

The term ψ in Eq. (3.29) can be expressed as

$$\psi = 1 + \psi_1 + \psi_2 + \psi_3 + \cdots$$

where ψ_1, ψ_2, ψ_3 can be expressed as

$$\psi_1 = \frac{\left(\frac{3}{2}+1\right)}{2}\xi, \quad \psi_2 = \frac{\left(\frac{3}{2}+2\right)}{3}\psi_1\xi, \quad \psi_3 = \frac{\left(\frac{3}{2}+3\right)}{4}\psi_2\xi$$

which facilitates recursive computation. Using this, the necessary state-dependent coefficient form can be written as

[2] The binomial expansion of $(1+x)^{-n}$ can be written as

$$(1+x)^{-n} = 1 - nx + \frac{n(n+1)}{2!}x^2 - \frac{n(n+1)(n+2)}{3!}x^3 + \cdots \tag{3.27}$$

$$\underbrace{\begin{bmatrix} \dot{x}_1 \\ \dot{x}_2 \\ \dot{x}_3 \\ \dot{x}_4 \\ \dot{x}_5 \\ \dot{x}_6 \end{bmatrix} = \begin{bmatrix} 0 & 1 & 0 & 0 & 0 & 0 \\ a_{21} & 0 & a_{23} & 2\dot{\nu}_c & a_{25} & 0 \\ 0 & 0 & 0 & 1 & 0 & 0 \\ -\ddot{\nu}_c & -2\dot{\nu}_c & \dot{\nu}_c^2 - \frac{\mu}{\gamma} & 0 & 0 & 0 \\ 0 & 0 & 0 & 0 & 0 & 1 \\ 0 & 0 & 0 & 0 & -\frac{\mu}{\gamma} & 1 \end{bmatrix}}_{A} \underbrace{\begin{bmatrix} x_1 \\ x_2 \\ x_3 \\ x_4 \\ x_5 \\ x_6 \end{bmatrix}}_{x} + \underbrace{\begin{bmatrix} 0&0&0 \\ 1&0&0 \\ 0&0&0 \\ 0&1&0 \\ 0&0&0 \\ 0&0&1 \end{bmatrix}}_{B} \boldsymbol{u}$$

where $a_{21} \triangleq \dot{\nu}_c^2 - \frac{\mu}{\gamma} + \frac{3\mu}{2r_c^3}\left(2 + \frac{x_1}{r_c}\right)\psi$, $a_{23} \triangleq \ddot{\nu}_c + \frac{3\mu}{2r_c^2}\psi x_3$, $a_{25} \triangleq \frac{3\mu}{2r_c^2}\psi x_5$.

3.5 Results and Discussions

In this section, comparative study of performance characteristics between LQR- and SDRE-based guidance has been presented based on infinite-time philosophy. For this analysis, weighting matrices on the state and control vectors are selected as $Q = I_6$ and $R = 10^9 I_3$, respectively. Note that infinite-time SDRE (denoted as I-SDRE in this book) are presented here, first using the SDC_1 formulation and subsequently using the SDC_2 formulation. Since the LQR technique works on the validity of the linear relative dynamics, which in turn relies on two assumptions, namely (i) the chief satellite is in the circular orbit and (ii) the relative distance between the deputy and chief satellite is small. Results are included in this section to both demonstrate this fact as well as to show that LQR philosophy does not perform satisfactorily in case either or both of these assumptions is/are not true. However, it is shown that SDRE performs better in comparison to the cases where the LQR fails.

Simulation results provided in this section can be obtained from the program files provided in the folder named '*Infinite-time: LQR and SDRE*'. For more details, refer to Section A.1.

3.5.1 Formation Flying in Circular Orbit and with Small Desired Relative Distance

The simulated case studies are so selected with initial condition and a desired final condition as given in Table 2.2 presented in Chapter 2. Note that, the choice of the final relative distance is small enough and the chief satellite is in the circular orbit so that the linear equation of motion is still valid.

The simulated relative trajectory of deputy satellite in chief-satellite-centered Hill's frame pertaining to formulations, viz., LQR, SDC_1, and SDC_2, are shown in Fig. 3.1. The deputy satellite starts from the inner initial relative formation trajectory and is commanded to the outer relative orbit. It can be easily verified that all formulation methods drive the deputy satellite to the desired relative states, since the underlying restrictive conditions that the chief satellite is in the circular orbit and the relative distance between the chief and the deputy is small is met in this case. Note that the LQR method which is purely based on the linear dynamic model also achieves desired relative states for this case. For better clarity, the formation reconfiguration trajectories are also shown in xy, xz, and yz planes in Figs. 3.2, 3.3, and 3.4, respectively. The state error history is plotted in Figs. 3.5 and 3.6, which shows the state errors reduce along the trajectory to achieve the desired condition. Figure 3.7 illustrates the optimal control corresponding to LQR, SDC_1, and SDC_2 required in achieving the desired state values $\mathbf{x}_d$. It is obvious that all control components are pretty close to each other. The terminal state errors achieved between LQR and both SDC models are tabulated in Table 3.1 and there is no appreciable difference in performance among the three formulations. This can be seen clearly from the computed norm of the state error shown in Table 3.1.

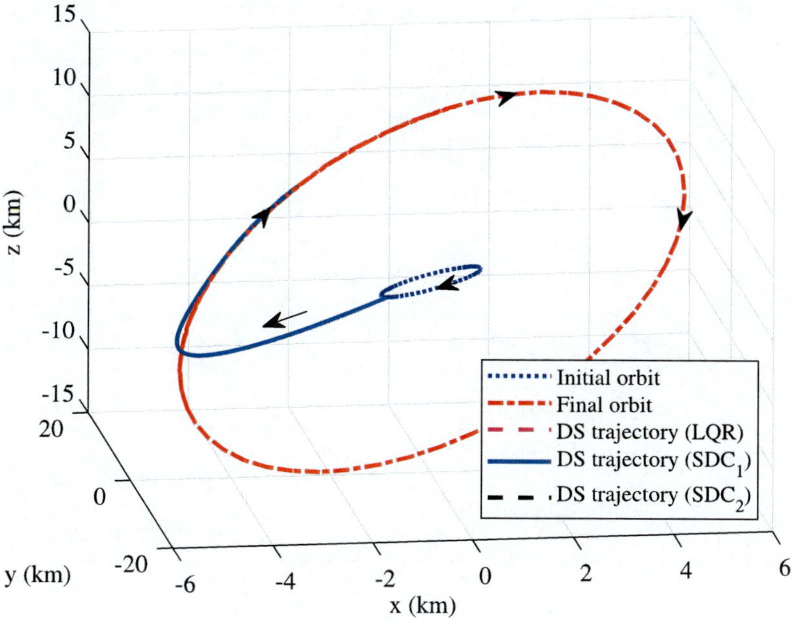

Fig. 3.1 Deputy satellite (DS) trajectory ($\rho_i = 1\,\text{km}$, $\rho_f = 5\,\text{km}$, $e = 0.0$)

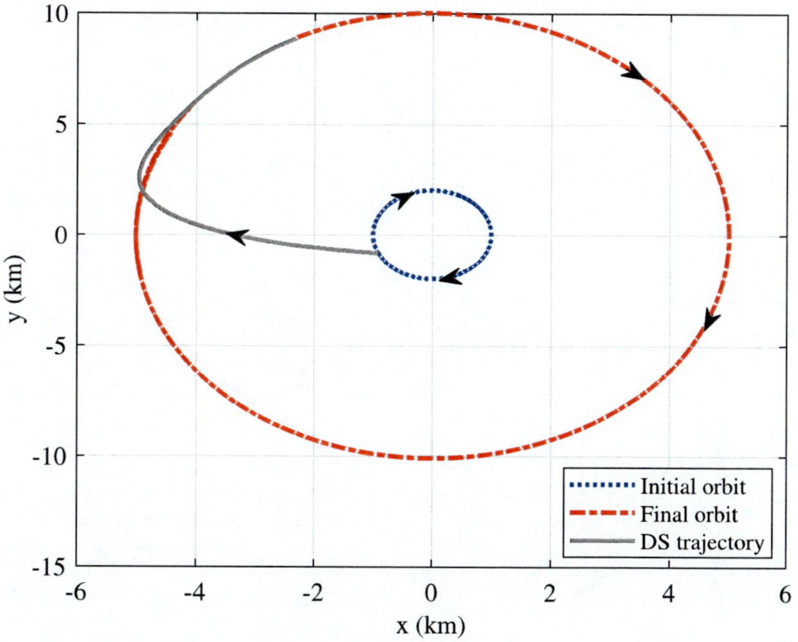

Fig. 3.2 Deputy satellite (DS) trajectory in xy plane of Hill's frame

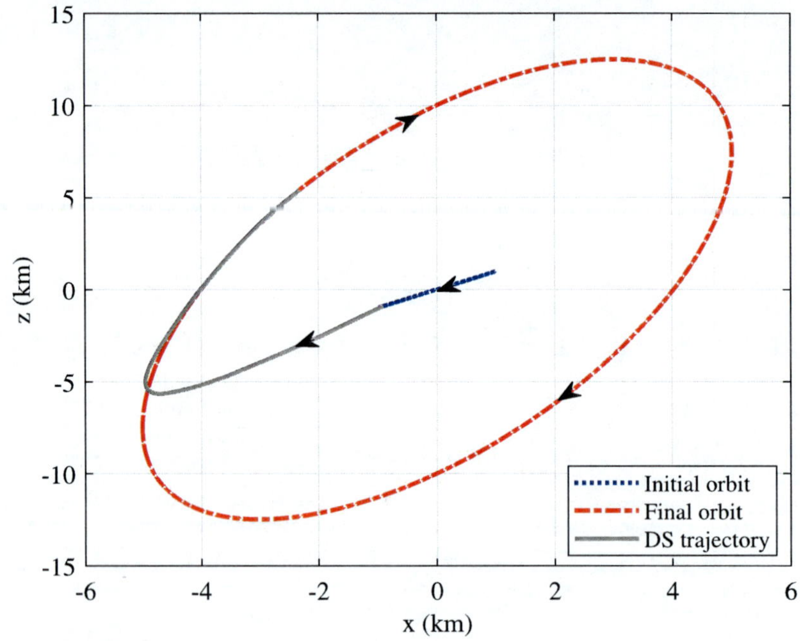

Fig. 3.3 Deputy satellite (DS) trajectory in xz plane of Hill's frame

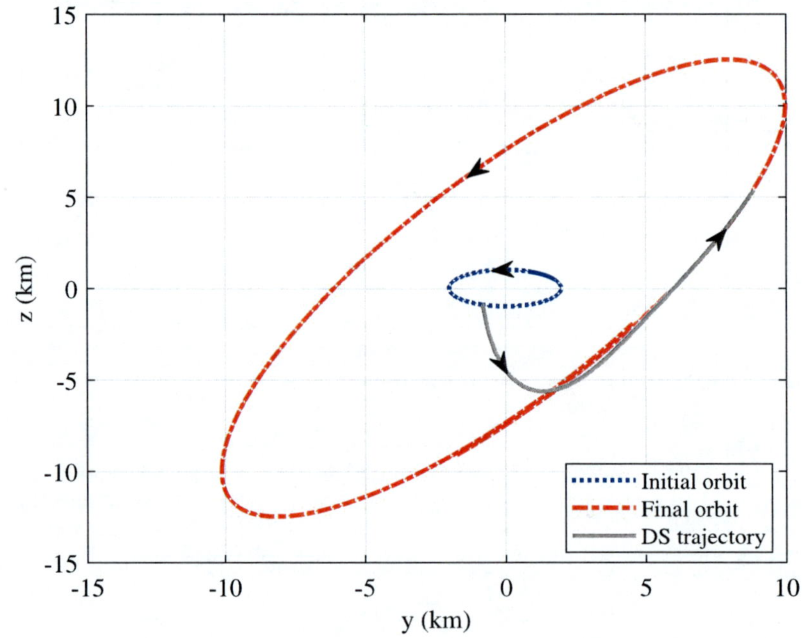

Fig. 3.4 Deputy satellite (DS) trajectory in yz plane

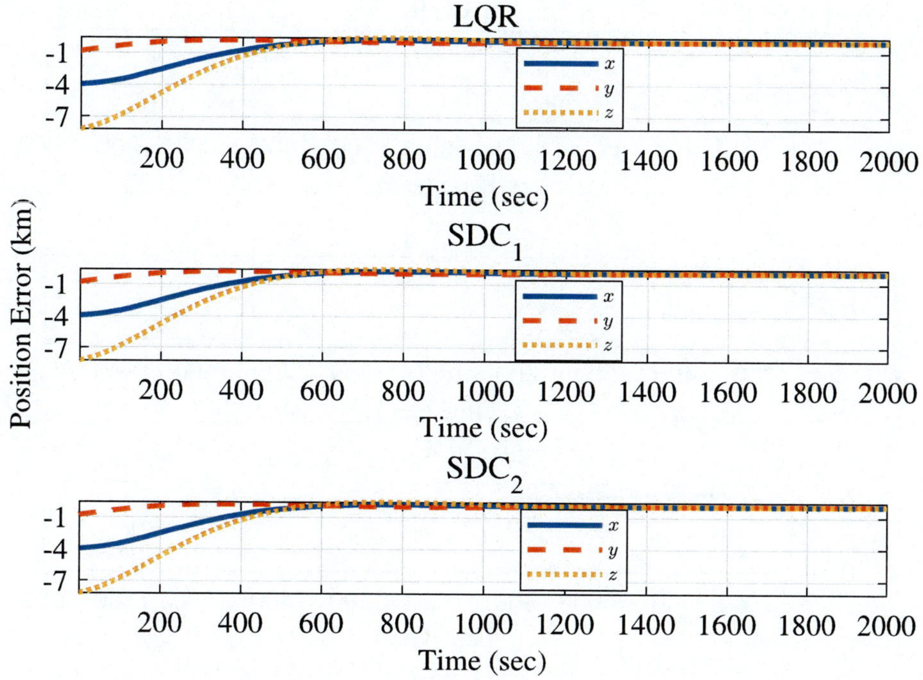

Fig. 3.5 Position error history ($\rho_i = 1$ km, $\rho_f = 5$ km, $e = 0.0$)

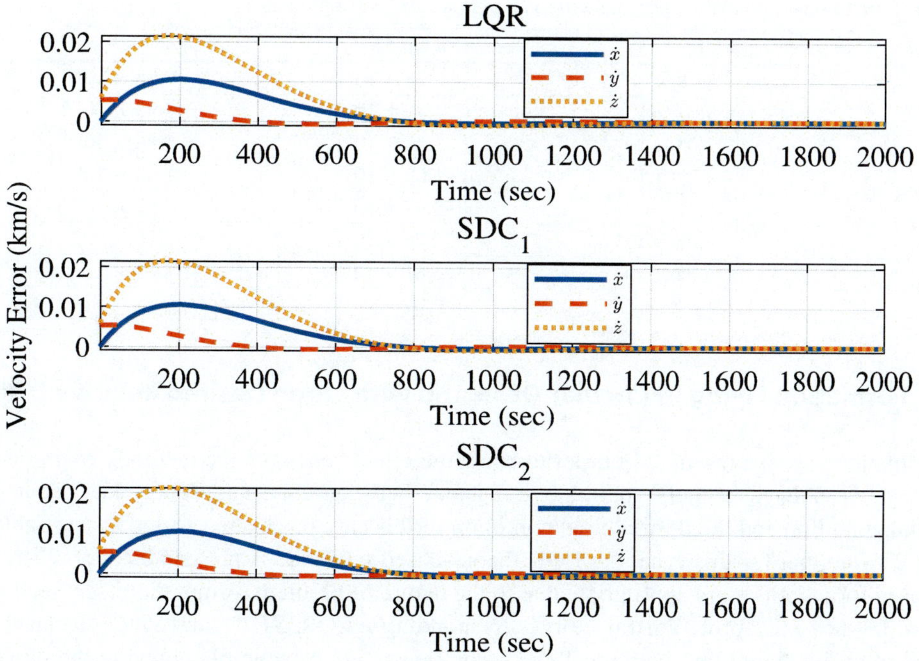

Fig. 3.6 Velocity error history ($\rho_i = 1$ km, $\rho_f = 5$ km, $e = 0.0$)

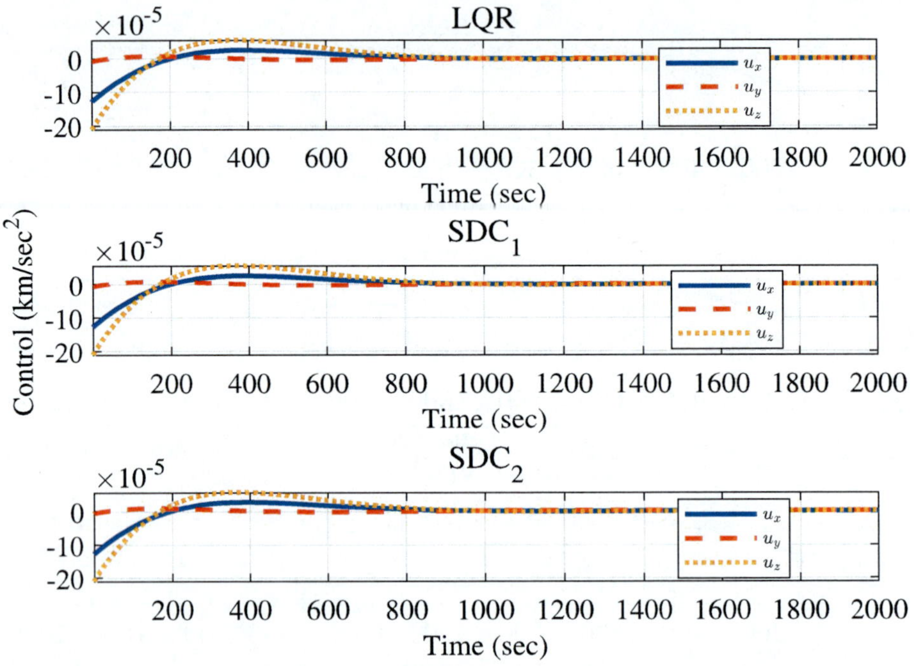

Fig. 3.7 Control history for deputy satellite ($\rho_i = 1$ km, $\rho_f = 5$ km, $e = 0.0$)

Table 3.1 Infinite-time formulation terminal error ($\rho_i = 1$ km, $\rho_f = 5$ km, $e = 0.0$)

State error	LQR	SDC_1	SDC_2
Δx (km)	8.48×10^{-4}	-9.28×10^{-4}	-9.28×10^{-4}
$\Delta \dot{x}$ (km/s)	9.67×10^{-6}	7.90×10^{-6}	7.90×10^{-6}
Δy (km)	3.05×10^{-3}	1.84×10^{-3}	1.84×10^{-3}
$\Delta \dot{y}$ (km/s)	-9.22×10^{-6}	-1.08×10^{-5}	-1.08×10^{-5}
Δz (km)	-5.47×10^{-3}	-2.97×10^{-3}	-2.97×10^{-3}
$\Delta \dot{z}$ (km/s)	2.68×10^{-5}	2.76×10^{-5}	2.76×10^{-5}
Norm	6.32×10^{-3}	3.61×10^{-3}	3.61×10^{-3}

3.5.2 Formation Flying in Circular Orbit and with Large Desired Relative Distance

This simulation case is presented in order to show the effectiveness of the methods when one of the assumptions is violated, namely, the relative distance between chief and deputy has become large. The initial condition and the desired final condition of this case has been selected as per Table 2.4 in Chapter 2. Note that for this case, the LQR theory-based guidance is not conductive, whereas both SDC based formulations are appropriate due to the nature of problem formulation (see Section 3.4).

Figure 3.8 shows the formation trajectory corresponding to LQR, SDC_1, and SDC_2. It can be clearly observed from this figure that both the SDC formulations are capable of achieving the objective of driving the deputy satellite to the desired configuration. Moreover, both formulations lead to the very same results with no appreciable difference between the two trajectories even under transient.

The position error evolution (Fig. 3.9) as well as velocity error evolution (Fig. 3.10) shows the state errors reduce along the trajectory to achieve the desired final states. The control solution corresponding to LQR, SDC_1, and SDC_2 are plotted in Fig. 3.11. The terminal state errors achieved by LQR and both SDC models are tabulated in Table 3.2. As expected, the results from the SDC model-based guidance strategies outperform the LQR theory-based guidance strategy. The final error in acquiring states for

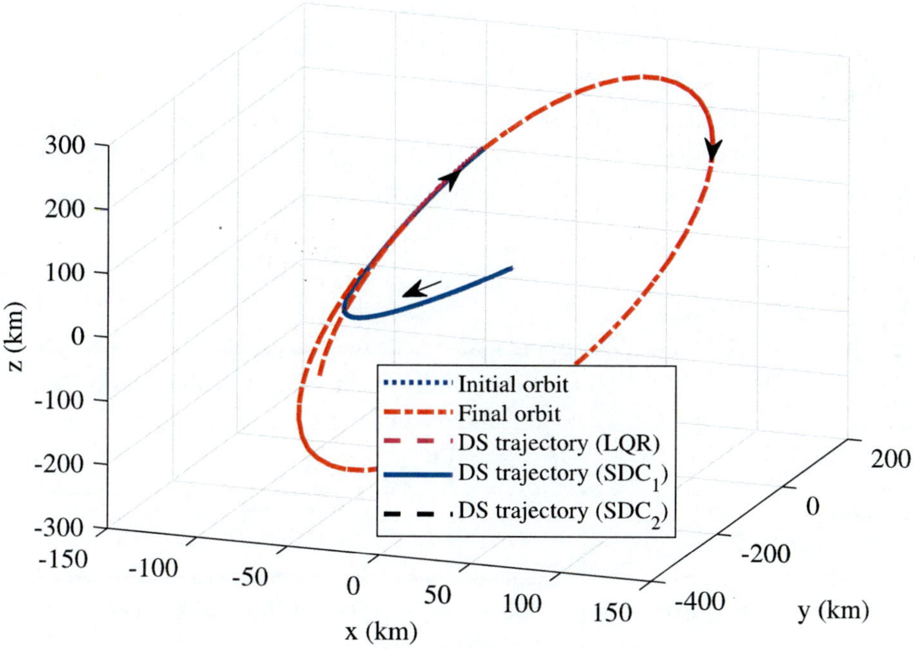

Fig. 3.8 Deputy satellite (DS) trajectory ($\rho_i = 1\,\text{km}$, $\rho_f = 100\,\text{km}$, $e = 0.0$)

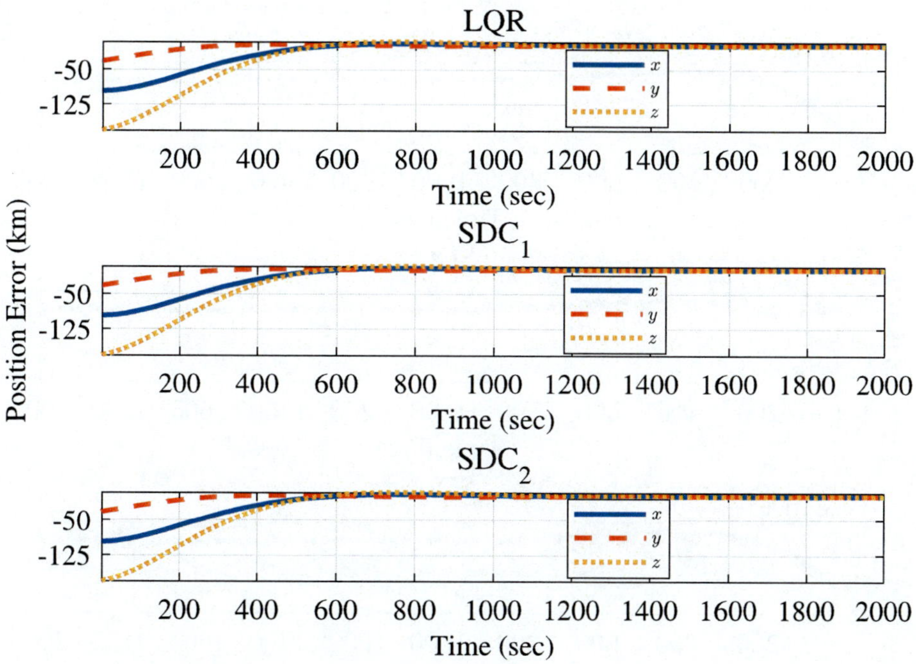

Fig. 3.9 Position error history ($\rho_i = 1\,\text{km}$, $\rho_f = 100\,\text{km}$, $e = 0.0$)

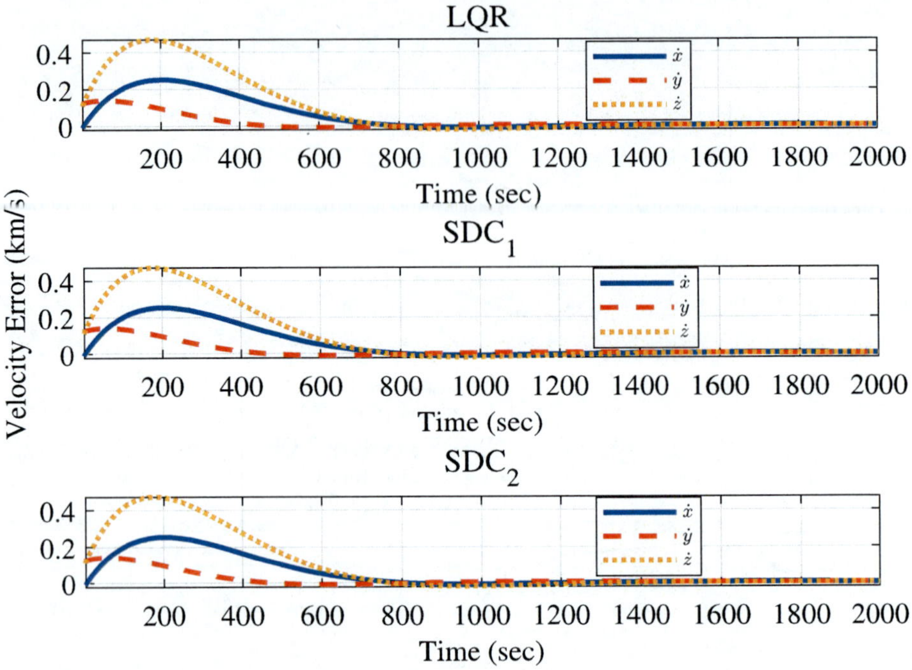

Fig. 3.10 Velocity error history ($\rho_i = 1\,\text{km}$, $\rho_f = 100\,\text{km}$, $e = 0.0$)

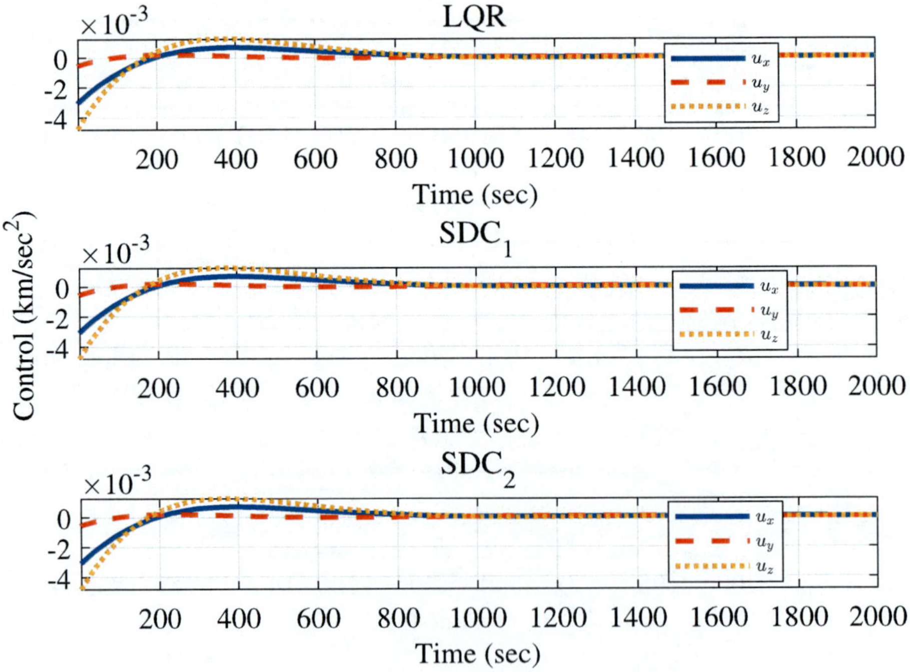

Fig. 3.11 Control history for deputy satellite ($\rho_i = 1\,\text{km}$, $\rho_f = 100\,\text{km}$, $e = 0.0$)

Table 3.2 Infinite-time formulation terminal error ($\rho_i = 1$ km, $\rho_f = 100$ km, $e = 0.0$)

State error	LQR	SDC_1	SDC_2
Δx (km)	6.96×10^{-1}	-2.35×10^{-2}	-2.35×10^{-2}
$\Delta \dot{x}$ (km/s)	9.33×10^{-4}	2.07×10^{-4}	2.07×10^{-4}
Δy (km)	5.61×10^{-1}	3.80×10^{-2}	3.80×10^{-2}
$\Delta \dot{y}$ (km/s)	3.93×10^{-4}	-2.37×10^{-4}	-2.37×10^{-4}
Δz (km)	-9.98×10^{-1}	-6.09×10^{-2}	-6.09×10^{-2}
$\Delta \dot{z}$ (km/s)	2.91×10^{-4}	5.85×10^{-4}	5.85×10^{-4}
Norm	1.34×10^{0}	7.56×10^{-2}	7.55×10^{-2}

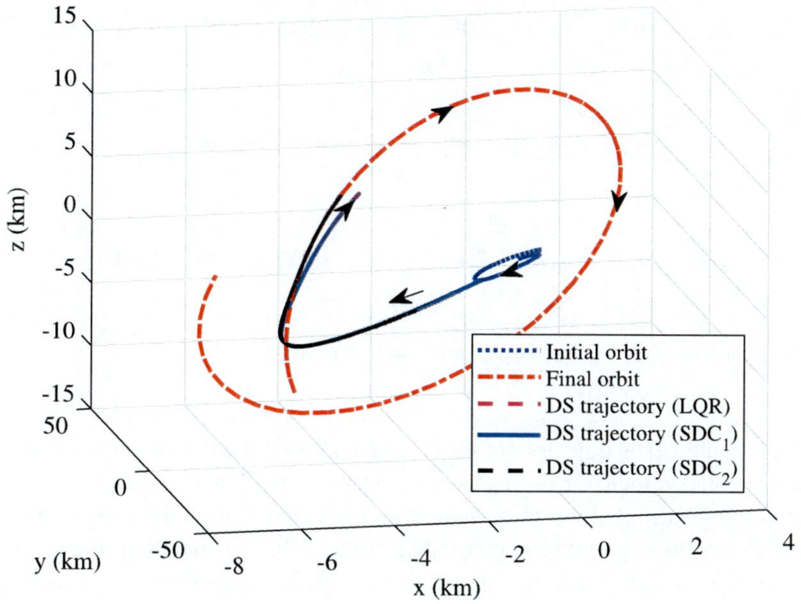

Fig. 3.12 Deputy satellite (DS) trajectory ($\rho_i = 1$ km, $\rho_f = 5$ km, $e = 0.05$)

the LQR guidance strategy has been observed to be two orders higher than the SDC-based models (see Table 3.2). Also, it is to be noted that both SDC-based formulations achieved terminal errors that are pretty close to each other because both models can handle the large desired relative distance.

3.5.3 Formation Flying in Elliptic Orbit and with Small Desired Relative Distance

In this case study, one of the main assumptions that the orbit of the chief satellite be circular is relaxed so that the orbit is slightly eccentric (not highly eccentric). The initial and final conditions for this case is selected from Table 2.3 of Chapter 2.

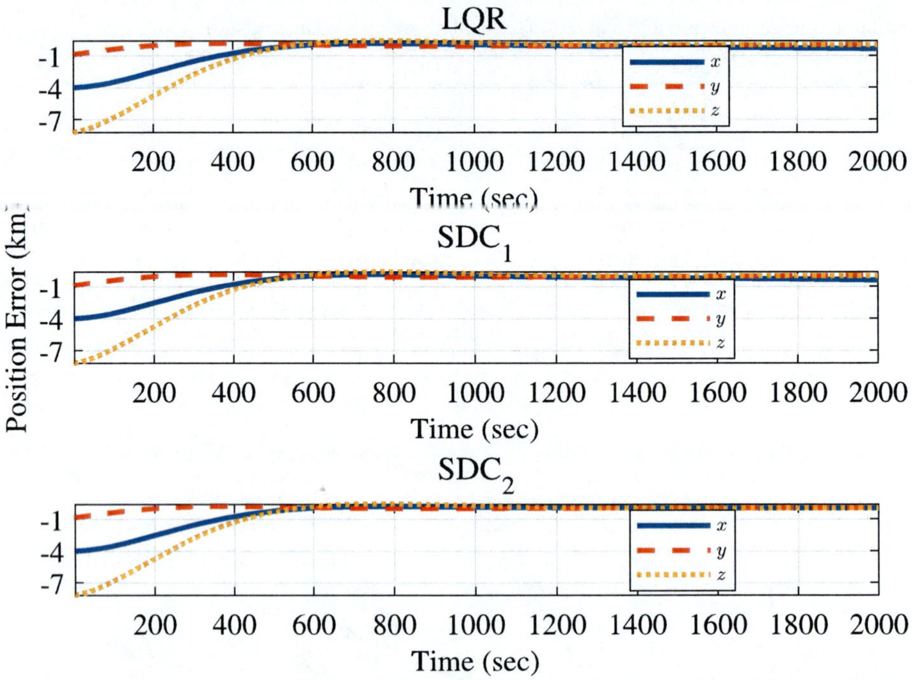

Fig. 3.13 Position error history ($\rho_i = 1$ km, $\rho_f = 5$ km, $e = 0.05$)

The simulation results of the trajectory of the deputy satellite in Hill's frame is presented in Fig. 3.12. The relative position and velocity error histories are provided in Figs. 3.13 and 3.14, respectively. The control history for the formulations is presented in Fig. 3.15. Notice the error norm computed in Table 3.3 clearly shows the superlative performance of the SDC_2 formulation. It can be observed that the state error based on LQR and SDC_1 model is much larger than the state error based on SDC_2 model since LQR and SDC_1 formulations are based on the assumption of the circular chief satellite orbit.

Table 3.3 Infinite-time formulation terminal error ($\rho_i = 1$ km, $\rho_f = 5$ km, $e = 0.05$)

State error	LQR	SDC_1	SDC_2
Δx (km)	-3.80×10^{-1}	-3.83×10^{-1}	-6.37×10^{-4}
$\Delta \dot{x}$ (km/s)	-3.69×10^{-4}	-3.72×10^{-4}	5.97×10^{-6}
Δy (km)	4.12×10^{-2}	4.01×10^{-2}	1.60×10^{-3}
$\Delta \dot{y}$ (km/s)	2.94×10^{-4}	2.92×10^{-4}	-1.10×10^{-5}
Δz (km)	-5.77×10^{-3}	-2.63×10^{-3}	-2.62×10^{-3}
$\Delta \dot{z}$ (km/s)	2.50×10^{-5}	2.59×10^{-5}	2.59×10^{-5}
Norm	3.82×10^{-1}	3.85×10^{-1}	3.13×10^{-3}

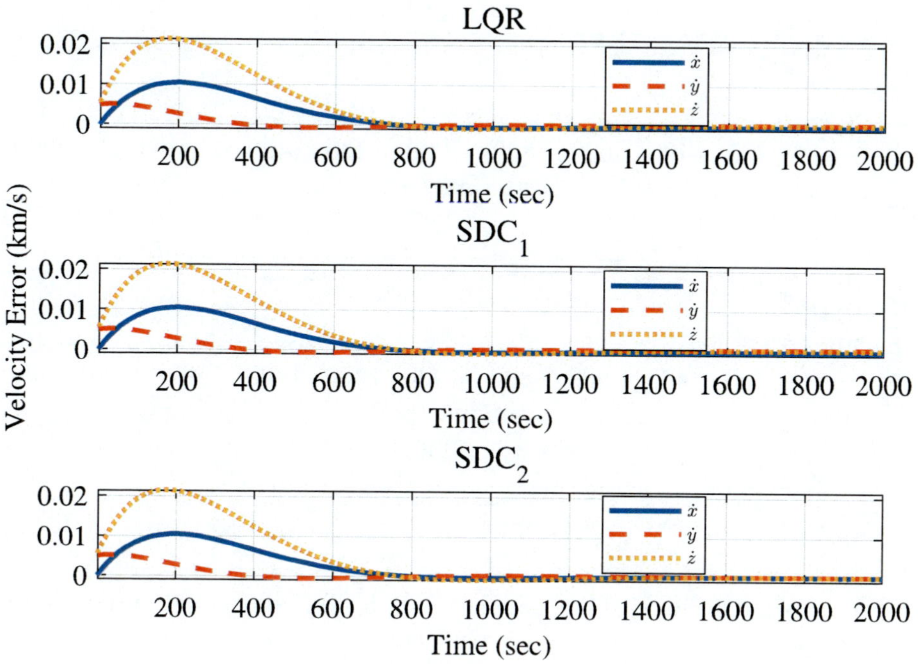

Fig. 3.14 Velocity error history ($\rho_i = 1$ km, $\rho_f = 5$ km, $e = 0.05$)

3.5.4 Formation Flying in Elliptic Orbit and with Large Desired Relative Distance

This specific simulation case is selected with initial condition and a desired final condition such that the chief orbit is elliptical and also the relative distance between the chief and deputy satellites is large. Table 2.5 of Chapter 2 presents the initial and final conditions for this simulation run.

Figure 3.16 clearly shows how large errors in the resulting desired configuration trajectory get corrected by SDC_2 more efficiently and effectively, whereas LQR and SDC_1 fail to do so. The position and velocity error histories are shown in Figs. 3.17 and 3.18, respectively. The corresponding control histories are depicted in Fig. 3.19. Once again, as expected, SDC_2 formulation performs better than the others as seen in Table 3.4 since the LQR and SDC_1 approximation of the actual plant is not holding good as the problem gets more nonlinear.

Table 3.4 Infinite-time formulation terminal error ($\rho_i = 1$ km, $\rho_f = 100$ km, $e = 0.05$)

State error	LQR	SDC_1	SDC_2
Δx (km)	-7.23×10^0	-8.35×10^0	-1.83×10^{-2}
$\Delta \dot{x}$ (km/s)	-7.18×10^{-3}	-8.56×10^{-3}	1.69×10^{-4}
Δy (km)	3.63×10^{-1}	-1.13×10^{-1}	3.49×10^{-2}
$\Delta \dot{y}$ (km/s)	6.52×10^{-3}	6.00×10^{-3}	-2.46×10^{-4}
Δz (km)	-1.23×10^0	-5.73×10^{-2}	-5.63×10^{-2}
$\Delta \dot{z}$ (km/s)	2.04×10^{-4}	5.41×10^{-4}	5.59×10^{-4}
Norm	7.35×10^0	8.35×10^0	6.87×10^{-2}

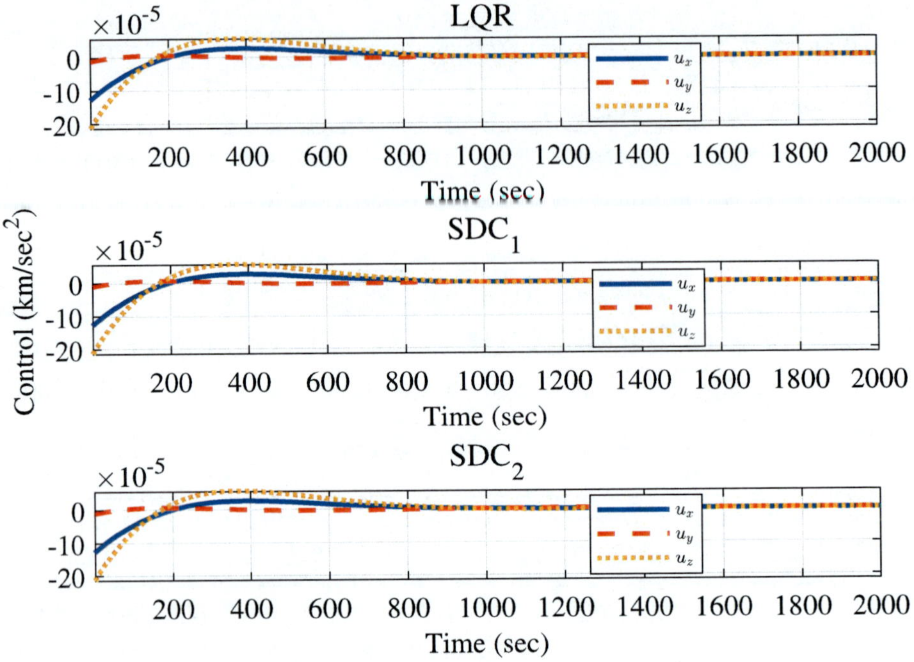

Fig. 3.15 Control history for deputy satellite ($\rho_i = 1$ km, $\rho_f = 5$ km, $e = 0.05$)

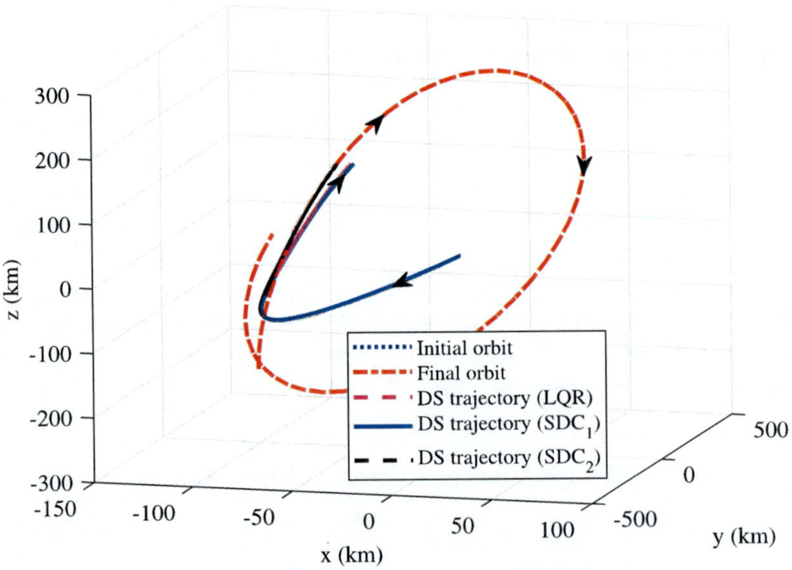

Fig. 3.16 Deputy satellite (DS) trajectory ($\rho_i = 1$ km, $\rho_f = 100$ km, $e = 0.05$)

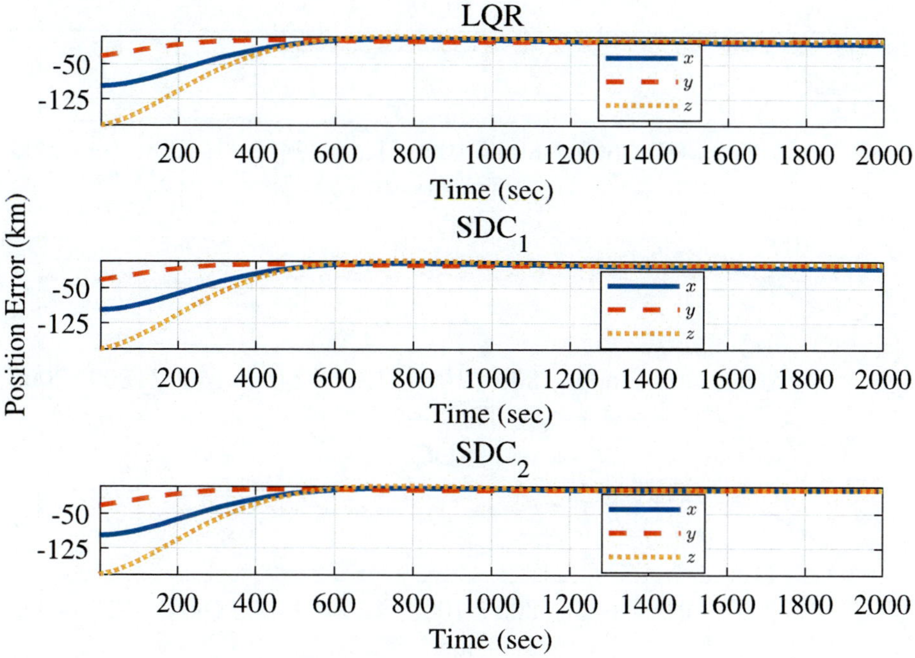

Fig. 3.17 Position error history ($\rho_i = 1$ km, $\rho_f = 100$ km, $e = 0.05$)

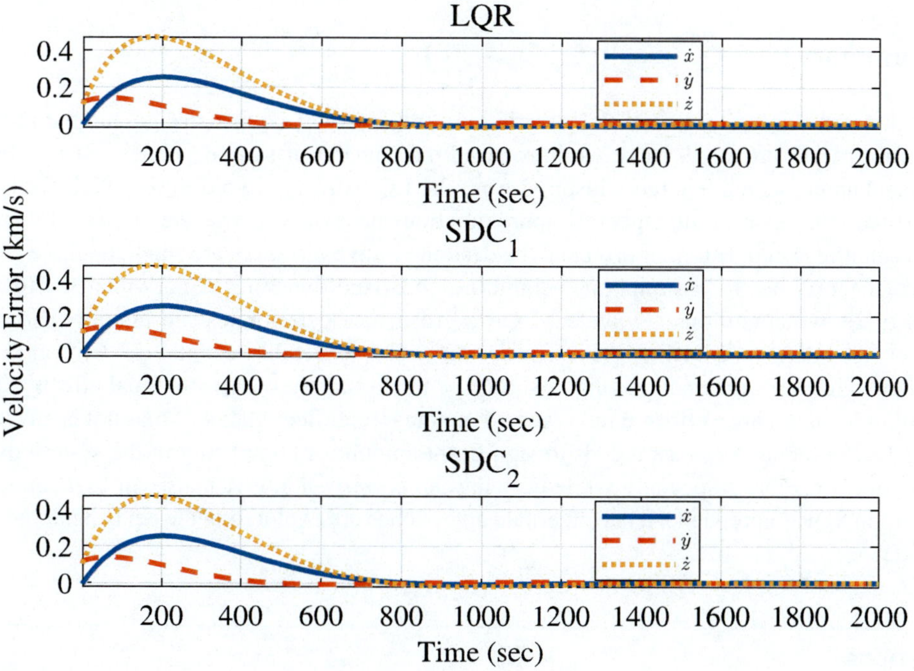

Fig. 3.18 Velocity error history ($\rho_i = 1$ km, $\rho_f = 100$ km, $e = 0.05$)

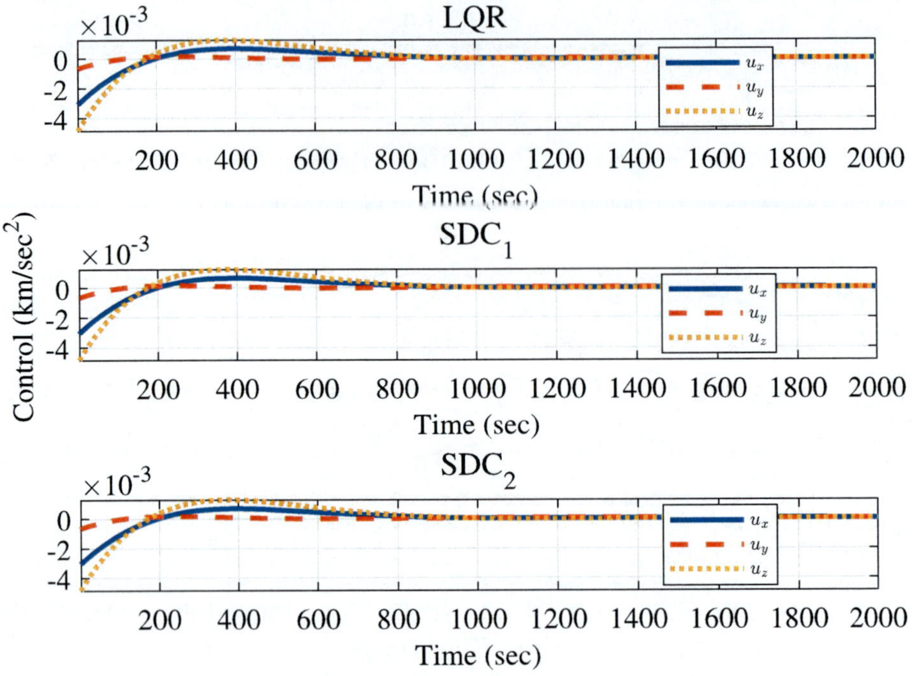

Fig. 3.19 Control history for DS ($\rho_i = 1$ km, $\rho_f = 100$ km, $e = 0.05$)

3.6 Summary

In this chapter, a brief overview of the generic infinite-time LQR control design and the basic computation steps were provided. Next, we introduced the concept of state-dependent Riccati Equation (SDRE) technique, which is largely inspired from the LQR theory. Two different SDC formulations are presented with appropriate algebraic approximations necessary. It was demonstrated that in case the chief satellite is in the circular orbit and the desired relative distances are small, all three controller works satisfactorily as the assumptions behind the linearized dynamics is not violated. However, if either of these two conditions is violated, LQR performance deteriorates substantially. However, in such cases, the nonlinear SDRE approach is recommended, preferably with the SDC_2 option.

Even though results are encouraging, it should be noted that the J_2 gravitational effect, which is a non-negligible disturbance effect in the system dynamics (refer Section 2.4.1), has not been taken into account. This is mainly because it adds to significant amount of complexity in the system dynamics and subsequent algebra, especially while deriving the corresponding A matrix in LQR approach or SDC form in SDRE approach. Hence, alternate approaches are explored in the next chapter to address this issue.

Y=N

References

1. Bryson, A.E., and Y.C. Ho. 1975. *Applied optimal control*. Hemisphere Publishing Corporation
2. Naidu, D.S. 2002. *Optimal control systems*. CRC Press.
3. Jin, X., and H. Lifu. 2011. *Formation keeping of micro-satellites LQR control algorithms analysis*, vol. 4.
4. Ogata, K., and Y. Yang. 2010. *Modern control engineering*, vol. 17. NJ: Pearson Upper Saddle River.

5. Nazarzadeh, J., M. Razzaghi, and K. Nikravesh. 1998. Solution of the matrix Riccati equation for the linear quadratic control problems. *Mathematical and Computer Modelling* 27 (7): 51–55.
6. Cloutier, J.R. 1997. State-dependent Riccati equation techniques: An overview. In *IEEE Proceedings of the 1997 American Control Conference*, vol. 2, 932–936
7. Cimen, T. 2008. State-dependent Riccati equation (SDRE) control: A survey. In *International federation of automatic control*, 3761–3775. Elsevier.
8. Irvin, D.J., and D.R. Jacques. 2002. A study of linear versus nonlinear control techniques for the reconfiguration of satellite formations. In *Advances in the astronautical sciences* 589–608.
9. Park, H.E., S.Y. Park, and K.H. Choi. 2011. Satellite formation reconfiguration and station keeping using SDRE technique. *Aerospace Science and Technology* 15: 440–452.
10. Kreyszig, E. 2009. *Advanced engineering mathematics*, 10th Edn. Technical report, Wiley.

Check for updates

4 Adaptive LQR for Satellite Formation Flying

As shown in Chapter 3, a linearized relative dynamics based LQR control theory does not lead to satisfactory results when the formation is desired in an elliptic orbit and/or when the relative separation is high. It was also demonstrated in Chapter 6 that the SDRE approach is a relatively better to address such issues to a reasonable extent. Unfortunately, however, practicing engineers are typically not comfortable to switch over to nonlinear control theory completely. In view of this, an alternate approach is presented here. This approach uses the standard infinite-time LQR control as the nominal controller. However, it is augmented with online learning based *adaptive optimal controller* that accounts for the effects of the neglected system dynamics and external state-dependent disturbances together. This additional adaptive component in fact is computed from the identified system. The overall controller is termed as '*Adaptive LQR*' controller for obvious reasons. As expected, this Adaptive LQR controller performs much better than the nominal LQR controller. It is worth mentioning here that this technique is capable enough to handle small eccentricity as well as large desired relative distance scenarios. Details of the process are discussed next.

4.1 Online Model Adaptation

The class of control-affine nonlinear dynamical systems considered here is assumed to satisfy the following system dynamics:

$$\dot{\mathbf{x}} = f(\mathbf{x}) + B\mathbf{u} + \mathbf{d}'(\mathbf{x}) \tag{4.1}$$

Here, apart from the standard meanings of $\mathbf{x}$, $\mathbf{u}$, f, and B (see Chapter 2), the notation $\mathbf{d}'(\mathbf{x})$ denotes the state-dependent external disturbance to the system, which is assumed to be unknown.

A key idea here is to first add and subtract $A\mathbf{x}$ and rewrite the system dynamics as

$$\dot{\mathbf{x}} = A\mathbf{x} + B\mathbf{u} + \mathbf{d}(\mathbf{x}) \tag{4.2}$$

where the matrix A is defined as

$$A = \left[\frac{\partial f(\mathbf{x})}{\partial \mathbf{x}}\right]_{\mathbf{x}_0} \tag{4.3}$$

Electronic supplementary material The online version of this chapter (https://doi.org/10.1007/978-981-15-9631-5_4) contains supplementary material, which is available to authorized users.

S. Mathavaraj and R. Padhi, *Satellite Formation Flying*,
https://doi.org/10.1007/978-981-15-9631-5_4

and $d(\mathbf{x}) = f(\mathbf{x}) - A\mathbf{x} + \mathbf{d}'(\mathbf{x})$ is assumed to be the total uncertainty term in the system dynamics. Note that $\mathbf{x}_0$ is the 'operating point' of the system.

For the satellite formation flying problem, as mentioned in Section 2.3.3, the linearization is carried out under the following assumptions: (i) circular reference orbit of the chief satellite around the Earth (ii) radial separation between the chief and deputy satellite is very small compared to the radius vector of the chief satellite from the center of Earth (iii) no external perturbations acting on the chief as well as deputy satellites. These assumptions result in a time-invariant A matrix for the system considered in Eq. (4.2).

The next key idea is to construct a 'disturbance observer', the dynamics of which is represented as

$$\dot{\mathbf{x}}_o = A\mathbf{x} + B\mathbf{u} + \hat{\mathbf{d}}(\mathbf{x}) + \mathbf{k}_o(\mathbf{x} - \mathbf{x}_o) \tag{4.4}$$

where $\hat{\mathbf{d}}(\mathbf{x})$ is an approximation of the actual function $\mathbf{d}(\mathbf{x})$ and $\mathbf{k}_o > 0$ is a positive definite gain matrix that needs to be selected by the designer (and tuned for good performance). The task here is to simultaneously ensure $\mathbf{x} \to \mathbf{x}_o$ and $\mathbf{d}(\mathbf{x}) \to \hat{\mathbf{d}}(\mathbf{x})$ as the plant starts operating. Note that once this objective is met, the identified observer dynamics can be a close representation of the actual system dynamics.

Since it is desired that the states of the observer plant should track the states of the actual plant, we can define the error in state as

$$\mathbf{e} \triangleq \mathbf{x} - \mathbf{x}_o \tag{4.5}$$

Next, the error dynamics is obtained by differentiating Eq. (4.5) and then substituting Eqs. (4.2) and (4.4), which results in

$$\begin{aligned}\dot{\mathbf{e}} &= \dot{\mathbf{x}} - \dot{\mathbf{x}}_o \\ &= [A\mathbf{x} + B\mathbf{u} + \mathbf{d}(\mathbf{x})] - \left[A\mathbf{x} + B\mathbf{u} + \hat{\mathbf{d}}(\mathbf{x}) + \mathbf{k}_o(\mathbf{x} - \mathbf{x}_o)\right] \\ &= \mathbf{d}(\mathbf{x}) - \hat{\mathbf{d}}(\mathbf{x}) - \mathbf{k}_o\mathbf{e}\end{aligned} \tag{4.6}$$

From Eq. (4.6), the ith channel ($i = 1, 2, \ldots, n$) error dynamics can be written as

$$\dot{e}_i = d_i(\mathbf{x}) - \hat{d}_i(\mathbf{x}) - k_{o_i} e_i \tag{4.7}$$

Next, the idea is to approximate $\hat{d}_i(\mathbf{x})$ as a linear-in-weight generic neural network and write it as

$$\hat{d}_i(\mathbf{x}) = \hat{W}_i^T \Phi_i(\mathbf{x}) \tag{4.8}$$

where $\hat{W}_i$ is the 'actual' weight vector of the neural network and $\Phi_i(\mathbf{x})$ is the associated basis functions. Even though ideally $\hat{W}_i$ is a 'constant' vector, it is assumed to be 'time-varying' so that it can be updated with the evolution of time, leading to an '*online training process*'. Most often it is found that all components of $\hat{W}_i$ evolve to steady-state (i.e., constant) values with the evolution of time.

Note that from the universal function approximation property of neural networks, it is known that there exists an ideal neural network with an optimal weight W_i and basis function $\Phi_i(\mathbf{x})$ that approximates $d_i(\mathbf{x})$ with accuracy ε_i. In other words, one can write

$$d_i(\mathbf{x}) = W_i^T \Phi_i(\mathbf{x}) + \varepsilon_i \tag{4.9}$$

The need for training the neural network arises because the ideal weight W_i is not known. In this development, it is assumed that both $W_i, \hat{W}_i \in \Re^{q_i \times 1}$, where q_i is the number of basis function considered to approximate the disturbance function $d_i(\mathbf{x})$ in the ith channel.

Next, the task is to propose a way of updating the weights of the neural network (i.e., training the network) such that the unknown function $\mathbf{d}(\mathbf{x})$ is captured. In this context, the channel-wise error dynamics Eq. (4.7) is rewritten as

$$\dot{e}_i = W_i{}^T \Phi_i(\mathbf{x}) + \varepsilon_i - \hat{W}_i{}^T \Phi_i(\mathbf{x}) - k_{o_i} e_i \tag{4.10}$$

Defining the error in weights of the ith network as $\tilde{W}_i \triangleq W_i - \hat{W}_i$, one can observe that $\dot{\tilde{W}}_i = -\dot{\hat{W}}_i$ as W_i is a constant weight vector. Next, the aim is to derive a weight update rule such that the weights of the approximating networks $\hat{W}_i$ approach the ideal weights W_i. To achieve this objective, a Lyapunov stability theory [1] based philosophy is followed, details of which are discussed next.

First, a positive definite Lyapunov function candidate is considered as

$$V_i(e_i, \tilde{W}_i) = \beta_i \frac{e_i^2}{2} + \frac{\tilde{W}_i{}^T \tilde{W}_i}{2\gamma_i} + \left[\frac{\partial d_i(\mathbf{x})}{\partial \mathbf{x}} - \frac{\partial \hat{d}_i(\mathbf{x})}{\partial \mathbf{x}}\right]^T \frac{\Theta_i}{2} \left[\frac{\partial d_i(\mathbf{x})}{\partial \mathbf{x}} - \frac{\partial \hat{d}_i(\mathbf{x})}{\partial \mathbf{x}}\right] \tag{4.11}$$

where β_i and γ_i are selected as positive constants and Θ_i is selected positive definite matrix. One can observe that the Lyapunov function V_i has the following three components:

(i) e_i, the error in states of ith channel,
(ii) $\tilde{W}_i$, the error in weights of ith network, and
(iii) $\left[\frac{\partial d_i(\mathbf{x})}{\partial \mathbf{x}} - \frac{\partial \hat{d}_i(\mathbf{x})}{\partial \mathbf{x}}\right]$, the error in partial derivative of ith unknown function.

It can be noted that ensuring the stability of the error dynamics in Eq. (4.10) for all $i = 1, \ldots, n$ ensures capturing of the disturbance function $d(\mathbf{x})$ through capturing the weights of the neural networks. In addition, it also ensures capturing the gradient of the disturbance functions in each channel, which leads to *directional learning* and hence minimization of transient effects.

Using Eqs. (4.8) and (4.9), one can rewrite Eq. (4.11) as

$$\begin{aligned} V_i(e_i, \tilde{W}_i) &= \beta_i \frac{e_i^2}{2} + \frac{\tilde{W}_i^T \tilde{W}_i}{2\gamma_i} + (W_i - \hat{W}_i)^T \left[\frac{\partial \Phi_i}{\partial \mathbf{x}}\right] \frac{\Theta_i}{2} \left[\frac{\partial \Phi_i}{\partial \mathbf{x}}\right]^T (W_i - \hat{W}_i) \\ &= \beta_i \frac{e_i^2}{2} + \frac{\tilde{W}_i{}^T \tilde{W}_i}{2\gamma_i} + \tilde{W}_i^T \left[\frac{\partial \Phi_i}{\partial \mathbf{x}}\right] \frac{\Theta_i}{2} \left[\frac{\partial \Phi_i}{\partial \mathbf{x}}\right]^T \tilde{W}_i \end{aligned} \tag{4.12}$$

The time derivative of Lyapunov function is derived as follows:

$$\dot{V}_i = \beta_i e_i \dot{e}_i - \frac{\tilde{W}_i^T \dot{\hat{W}}_i}{\gamma_i} - \tilde{W}_i \left[\frac{\partial \Phi_i}{\partial \mathbf{x}}\right] \Theta_i \left[\frac{\partial \Phi_i}{\partial \mathbf{x}}\right]^T \dot{\hat{W}}_i \tag{4.13}$$

where higher order partial derivative terms are assumed to be zero. Substituting the error dynamics from Eq. (4.10) in Eq. (4.13) leads to

$$\dot{V}_i = \beta_i e_i \{\tilde{W}_i^T \Phi_i(\mathbf{x}) + \varepsilon_i - k_{o_i} e_i\} - \frac{\tilde{W}_i^T \dot{\hat{W}}_i}{\gamma_i} - \tilde{W}_i \left[\frac{\partial \Phi_i}{\partial \mathbf{x}}\right] \Theta_i \left[\frac{\partial \Phi_i}{\partial \mathbf{x}}\right]^T \dot{\hat{W}}_i \tag{4.14}$$

However, one can notice here that $\tilde{W}$ is unknown and hence further analysis of $\dot{V}_i$ is difficult. To proceed with the analysis avoiding this difficulty, the terms multiplying $\tilde{W}$ in Eq. (4.14) is forced to zero.

$$\left(\frac{I_{qi}}{\gamma_i} + \left[\frac{\partial \Phi_i}{\partial \mathbf{x}}\right] \Theta_i \left[\frac{\partial \Phi_i}{\partial \mathbf{x}}\right]^T\right) \dot{\hat{W}}_i = \beta_i e_i \Phi_i(\mathbf{x}) \tag{4.15}$$

where I_{q_i} is the identity matrix of dimension $q_i \times q_i$. Finally, from Eq. (4.15), the weight update rule can be written as

$$\dot{\hat{W}}_i = \beta_i e_i \left(\frac{I_p}{\gamma_i} + \left[\frac{\partial \Phi_i}{\partial \mathbf{x}}\right] \Theta_i \left[\frac{\partial \Phi_i}{\partial \mathbf{x}}\right]^T\right)^{-1} \Phi_i(\mathbf{x}) \tag{4.16}$$

With Eq. (4.16) in place, Eq. (4.14) can be written as

$$\dot{V}_i = \beta_i e_i \varepsilon_i - k_{o_i} \beta_i e_i^2 \tag{4.17}$$

From Eq. (4.17), it is obvious that $\dot{V}_i < 0$ provided

$$|e_i| > \frac{|\varepsilon_i|}{k_{o_i}} \tag{4.18}$$

This conditional stability is often referred to as '*practical stability*' in the literature [2]. From Eq. (4.18), one can conclude that if the absolute error in the ith channel exceeds the value $|\varepsilon_i|/k_{o_i}$, then $\dot{V}_i$ becomes negative. Hence, one can conclude that the trajectory of e_i will be pulled toward the bound defined in Eq. (4.18). In other words, the dynamics of the state error remains '*practically stable*', i.e., as soon as the error bound is violated, it will be pulled toward the bound. It is important to note here that irrespective of the function approximation error $|\varepsilon_i|$, the bound condition can be made smaller by increasing the gain value k_{o_i}, which is a design parameter. One can find this approach well documented in [3]. Note that even though the Lyapunov function has a gradient term in it, i.e., a directional learning approach is followed here which facilitates faster learning with lesser transient, like any other learning process, it still introduces a transient process. To minimize its effect on the synthesis of the additional controller which is discussed next, it is advisable to run the online adaptation process for a small finite-time window prior to this process. In fact, the derivation of the following section assumes that the online training process is over, i.e., weights have already attained the steady-state values.

It can be mentioned here that after the transient process is over, $\hat{\mathbf{d}}(\mathbf{x})$ becomes a close approximation of $\mathbf{d}(\mathbf{x})$. This results in better modeling of the system dynamics, which in turn facilitates synthesis of an additional optimal controller. Details of this process is described in the following section.

4.2 Adaptive LQR Theory

With the assumption that the function learning happens fast, i.e., $\hat{\mathbf{d}}(\mathbf{x}) \to \mathbf{d}(\mathbf{x})$ quickly in the close neighborhood of $\mathbf{d}(\mathbf{x})$ and remains so for all future time, the following representation of 'approximate system dynamics' can be constructed online by replacing $\mathbf{d}(\mathbf{x})$ in Eq. (4.2) by $\hat{\mathbf{d}}(\mathbf{x})$

$$\dot{\mathbf{x}} = A\mathbf{x} + B\mathbf{u} + \hat{\mathbf{d}}(\mathbf{x}) \tag{4.19}$$

Note that for simplicity a slight abuse of nomenclature and notation has been adopted here by defining Eq. (4.19) as 'approximate system dynamics'. This should not be confused with the definitions of the state $\mathbf{x}$ and control $\mathbf{u}$ of the actual system Eq. (4.2). It can be mentioned here that $\hat{\mathbf{d}}(\mathbf{x}) \triangleq \left[\hat{d}_1(\mathbf{x}), \ldots, \hat{d}_n(\mathbf{x})\right]^T$, where $\hat{d}_i(\mathbf{x}) = \hat{\mathbf{W}}_i^T \Phi_i(\mathbf{x})$ is a neural network represented as NN_2 in Section 4.2.1. Note that there can be situations, where $\hat{d}_i(\mathbf{x}) = 0$ for a few channels (such as kinematic part of the system dynamics). In that case, in those channels the function components need not be learned.

The objective is to drive the state to zero, which can be done by minimizing the following cost function:

$$J = \frac{1}{2} \int_{t_0}^{\infty} \left(\mathbf{x}^T Q \mathbf{x} + \mathbf{u}^T R \mathbf{u}\right) dt \tag{4.20}$$

where Q and R are positive semi-definite and positive definite state and control weighting matrices, respectively, which needs to be selected by a control designer judiciously.

The problem now is to minimize the cost function in Eq. (4.20) subject to the constraint in Eq. (4.19). However, since Eq. (4.19) represents a nonlinear system dynamics, the necessary conditions of optimality arising out of the nonlinear optimal control theory [4] needs to be manipulated, which are as follows:

$$\dot{\mathbf{x}} = A\mathbf{x} + B\mathbf{u} + \hat{\mathbf{d}}(\mathbf{x}) \tag{4.21}$$

$$\dot{\boldsymbol{\lambda}} = -Q\mathbf{x} - A^T \boldsymbol{\lambda} - \frac{\partial \hat{\mathbf{d}}}{\partial \mathbf{x}} \tag{4.22}$$

$$\mathbf{u} = -R^{-1} B^T \boldsymbol{\lambda} \tag{4.23}$$

In this approach, feed-forward neural networks are used next to augment the LQR controller to propose the Adaptive LQR philosophy in Section 4.2.1. Such networks are, however, trained using standard static optimization theory [5]. To facilitate that, discussion in rest of this chapter is done in the discrete-time framework. Discretizing state Eq. (4.21) using forward Euler integration and costate Eq. (4.22) using backward Euler integration, with a step size Δt, the necessary conditions of optimality can be written as

$$\begin{aligned} \mathbf{x}_{k+1} &= \mathbf{x}_k + \Delta t \left[A\mathbf{x}_k + B\mathbf{u}_k + \hat{\mathbf{d}}(\mathbf{x}_k)\right] \\ &= A_k \mathbf{x}_k + B_k \mathbf{u}_k + \hat{\mathbf{d}}_k(\mathbf{x}_k) \end{aligned} \tag{4.24}$$

$$\begin{aligned} \boldsymbol{\lambda}_k &= \boldsymbol{\lambda}_{k+1} - \Delta t \left[-Q\mathbf{x}_k - A^T \boldsymbol{\lambda}_{k+1} - \frac{\partial \hat{\mathbf{d}}}{\partial \mathbf{x}_k}\right] \\ &= Q_k \mathbf{x}_k + A_k^T \boldsymbol{\lambda}_{k+1} + \frac{\partial \hat{\mathbf{d}}_k}{\partial \mathbf{x}_k} \end{aligned} \tag{4.25}$$

$$\mathbf{u}_k = -R_k^{-1} B_k^T \boldsymbol{\lambda}_{k+1} \tag{4.26}$$

where $A_k = (I + A\,\Delta t)$, $B_k = B\,\Delta t$ and $\hat{\mathbf{d}}_k = \hat{\mathbf{d}}\,\Delta t$ are the discretized system matrix, control influence matrix, and total uncertainty term in the system, respectively. Also, the matrices $Q_k = Q\,\Delta t$ and $R_k = R\,\Delta t$ are positive semi-definite and positive definite state and control weighting matrices. Note that $\boldsymbol{\lambda}_k$ expression in Eq. (4.25) contains the expression $[\partial \hat{\mathbf{d}}_k / \partial \mathbf{x}_k]$. In order to evaluate this, the disturbance function learning process, as discussed in Section 4.1, also continues online. Hence, the disturbance observer as proposed in Eq. (4.4) and the associated neural network training process as proposed in Eq. (4.16) also play important roles in the overall synthesis of the Adaptive LQR controller.

4.2.1 Adaptive LQR Synthesis

First, one can observe that $\boldsymbol{\lambda}_{k+1}$ for the approximate system dynamics (4.19) is different from that of the standard linear quadratic regulator costate dynamics [6] because of the additional term $[\partial \hat{\mathbf{d}}_k / \partial \mathbf{x}_k]$ in Eq. (4.25). In view of this, the costate variable for the approximate system at t_{k+1} is first written as $\boldsymbol{\lambda}_{k+1} = \boldsymbol{\lambda}_{1_{k+1}} + \boldsymbol{\lambda}_{2_{k+1}}$, in which the costate variable $\boldsymbol{\lambda}_{1_{k+1}}$ is evaluated using the standard LQR results, i.e.,

$$\boldsymbol{\lambda}_{1_{k+1}} = P\mathbf{x}_k \tag{4.27}$$

where P is the Riccati matrix [6] obtained from the solution of the Riccati equation (3.7).

Next, $\boldsymbol{\lambda}_{2_{k+1}}$ is the additional costate component, which is necessary because of the modified system dynamics. The idea is to generate this $\boldsymbol{\lambda}_{2_{k+1}}$ from a neural network, denoted as NN_1, which needs to be trained online.

Taking into account the fact that in the process the function learning should also capture the unmodeled part of the system dynamics and hence the Adaptive LQR philosophy relies on the synthesis of two neural networks, named as NN_1 and NN_2. The overall philosophy is depicted in Fig. 4.1, where both NN_1 and NN_2 are trained online.

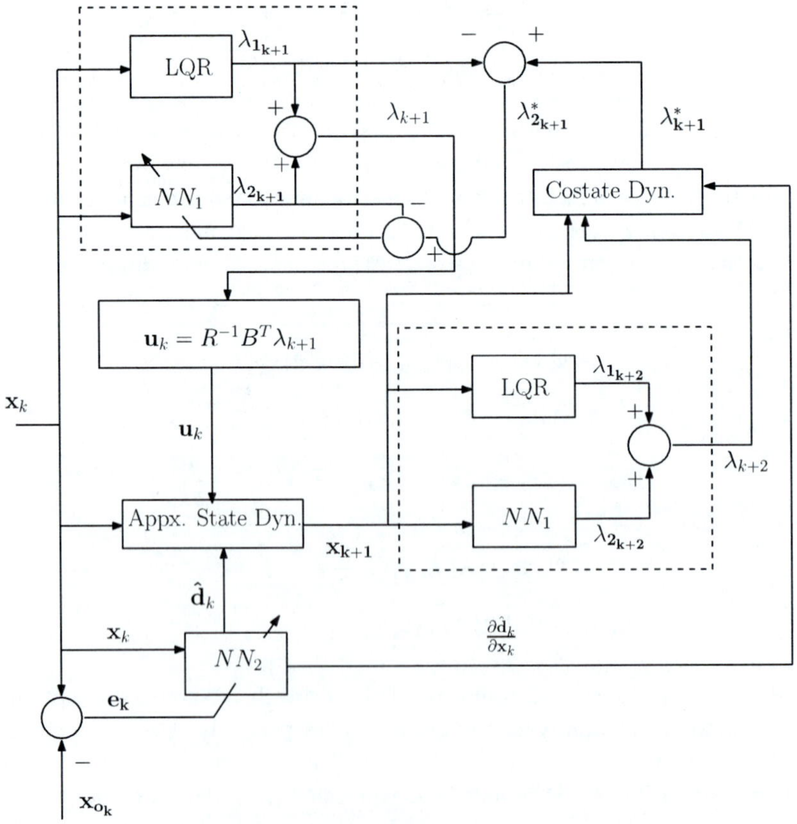

Fig. 4.1 Neural network scheme for Adaptive LQR design

Neural network NN_2 is trained in such a manner that both the unmodeled part of the dynamics as well as its partial derivatives with respect to the states are captured. Neural network NN_1 is trained in parallel so that an additional component of costate is captured. Note that the Adaptive LQR structure essentially adapts the offline optimal LQR controller for the nominal plant toward the optimal controller of the actual plant. The process can be summarized in three major blocks, which can be described as follows.

- LQR: This is the LQR block, which outputs the costate $\boldsymbol{\lambda}_{1_{k+1}} = P\mathbf{x}_k$ based on offline optimal discrete-time LQR problem. Hence, the derived costate value $\boldsymbol{\lambda}_{1_{k+1}}$ does not account for the unknown disturbance term $\mathbf{d}(\mathbf{x})$.
- NN_2: This neural network approximates the unmodeled dynamics which is crucial for online training. The weights use the channel-wise error information in the state for training. These are single-layer linear-in-the-weight networks based on radial basis functions. Training of NN_2 is done following the procedure outlined in Section 4.1, particularly using weight update rule Eq. (4.16) with zero initial condition.
- NN_1: This neural network approximates the additional costate required based on the information given by the NN_2 online training algorithm. Typically a linear-in-the-weight neural network, such as a radial basis function neural network, is used here. Training of NN_1 is described below in Section 4.2.2.

4.2.2 Synthesis of NN_1 Neural Network

The synthesis procedure of the NN_1 neural network, which is a key component of the adaptive architecture (see Fig. 4.1), is summarized below in two parts, i.e., (i) data generation procedure and (ii) NN_1 weight update rule.

4.2.2.1 Data Generation Procedure

The data generation procedure is done by following the steps mentioned below:

(i) Initialize NN_1 with zero weights.
(ii) Starting from the initial condition, input $\mathbf{x}_k$ (state of the actual system at time t_k) to both LQR and NN_2 block to obtain $\boldsymbol{\lambda}_{1_{k+1}}$ and $\boldsymbol{\lambda}_{2_{k+1}}$, respectively. Construct $\boldsymbol{\lambda}_{k+1} = \boldsymbol{\lambda}_{1_{k+1}} + \boldsymbol{\lambda}_{2_{k+1}}$.
(iii) Use $\mathbf{x}_k$ and $\boldsymbol{\lambda}_{k+1}$ in the optimal control Eq. (4.26) to get $\mathbf{u}_k$.
(iv) Use $\mathbf{x}_k$ and $\mathbf{u}_k$, as well as $\hat{\mathbf{d}}(\mathbf{x}_k)$ from NN_2, in the representation of the approximate state Eq. (4.24) to predict $\mathbf{x}_{k+1}$.
(v) Input $\mathbf{x}_{k+1}$ to both LQR and NN_1 again to obtain $\boldsymbol{\lambda}_{1_{k+2}}$ and $\boldsymbol{\lambda}_{2_{k+2}}$, respectively. Construct $\boldsymbol{\lambda}_{k+2} = \boldsymbol{\lambda}_{1_{k+2}} + \boldsymbol{\lambda}_{2_{k+2}}$.
(vi) Use $\mathbf{x}_{k+1}, \boldsymbol{\lambda}_{k+2}$ and $[\partial\hat{\mathbf{d}}_k/\partial\mathbf{x}_k]$ from NN_2 in the costate Eq. (4.25) (a backward propagation equation) to compute the 'target' $\boldsymbol{\lambda}^t_{k+1}$.
(vii) Construct $\boldsymbol{\lambda}^t_{2_{k+1}} = \boldsymbol{\lambda}^t_{k+1} - \boldsymbol{\lambda}_{1_{k+1}}$.
(viii) Adjust the weights of NN_1 such that the error between $\boldsymbol{\lambda}^t_{2_{k+1}}$ and $\boldsymbol{\lambda}_{2_{k+1}}$ is minimum.

4.2.2.2 NN_1 Weight Update Rule

The NN_1 neural network is typically constructed as a radial basis function network which is linear in the weight. Hence it can be represented as $\boldsymbol{\lambda}_{2_{k+1}} = W^T_{c_k}\phi_c(\mathbf{x}_k)$, where W_{c_k} is the current weight vector and $\phi_c(\mathbf{x}_k)$ is the vector containing the selected basis functions. The weight update rule for NN_1 network is derived by minimizing the following static cost function at kth instant of time:

$$J_{NN_1} = \frac{1}{2}\left(W_{ck}^{*T}\phi_c(\mathbf{x}_k) - W_{ck}^T\phi_c(\mathbf{x}_k)\right)^T\left(W_{ck}^{*T}\phi_c(\mathbf{x}_k) - W_{ck}^T\phi_c(\mathbf{x}_k)\right)$$
$$+\frac{1}{2}\left(W_{c_k} - W_{ck}^P\right)^T Q_w \left(W_{c_k} - W_{ck}^P\right) \tag{4.28}$$

where W_{ck}^{*T} is the target weight vector, Q_w is the weight on error term $\left(W_{c_k} - W_{ck}^P\right)$. The first term in Eq. (4.28) is included to minimize the error between the target $\boldsymbol{\lambda}_{2k+1}^t = W_{ck}^{*T}\phi_c(\mathbf{x}_k)$ and the actual $\boldsymbol{\lambda}_{2k+1} = W_{ck}^T\phi_c(\mathbf{x}_k)$, which is the primary objective. The second term, however, ensures that the weight W_{c_k} does not deviate too much from a previous trained value W_{ck}^P in this update process. This is included so that there is no sharp variation of the weights leading to smooth control action.

Next, the cost function in Eq. (4.28) is optimized in a 'free static optimization' process. Following the standard procedure [4], the necessary condition of optimality is given by

$$\frac{\partial J_{NN_1}}{\partial W_{c_k}} = \left(W_{ck}^{*T}\phi_c(\mathbf{x}_k)\phi_c^T(\mathbf{x}_k) + W_{ck}^{P\,T}Q_w\right) - W_{c_k}\left(\phi_c^T(\mathbf{x}_k)\phi_c(\mathbf{x}_k) + Q_w I\right)$$
$$= 0 \tag{4.29}$$

From Eq. (4.29), the expression for W_{c_k} can be obtained as

$$W_{c_k} = \frac{\left(W_{ck}^{*T}\phi_c(\mathbf{x}_k)\phi_c^T(\mathbf{x}_k) + W_{ck}^{P\,T}Q_w\right)}{\left(\phi_c^T(\mathbf{x}_k)\phi_c(\mathbf{x}_k) + Q_w I\right)} \tag{4.30}$$

where I is identity matrix of size $\phi_{c_k}\phi_{c_k}^T$.

The basis functions can either be chosen from a careful analysis of system dynamics for structured uncertainties (such as linearly appearing parameters) or selected as a set of generic functions (such as Gaussian functions) for unstructured uncertainties. For satellite formation flying problem, in each channel, the Gaussian basis function vector is selected as defined below.

$$\phi_c(\mathbf{x}_k) = e^{-\frac{1}{2}\left[\frac{(x_k-\mu_k)^2}{\sigma_k^2}\right]} \tag{4.31}$$

where μ_k, σ_k are the mean and variance of the actual system, respectively.

4.3 SFF Problem Formulation in Adaptive LQR Framework

The nonlinear relative dynamics between the deputy and chief satellite including J_2 perturbation can be written as

$$\underbrace{\begin{bmatrix}\dot{x}_1\\ \dot{x}_2\\ \dot{x}_3\\ \dot{x}_4\\ \dot{x}_5\\ \dot{x}_6\end{bmatrix}}_{\dot{\mathbf{x}}} = \underbrace{\begin{bmatrix} x_2 \\ 2\dot{\nu}_c x_4 + \ddot{\nu}_c x_3 + \dot{\nu}_c^2 x_1 - \frac{\mu}{\gamma}(x_1 + r_c) + \frac{\mu}{r_c^2} \\ x_4 \\ -2\dot{\nu}_c x_2 - \ddot{\nu}_c x_1 + \dot{\nu}_c^2 x_3 - \frac{\mu}{\gamma}x_3 \\ x_6 \\ -\frac{\mu}{\gamma}x_5 \end{bmatrix}}_{f(\mathbf{x})} + \underbrace{\begin{bmatrix} 0\,0\,0\\ 1\,0\,0\\ 0\,0\,0\\ 0\,1\,0\\ 0\,0\,0\\ 0\,0\,1\end{bmatrix}}_{B}\mathbf{u} + \underbrace{\begin{bmatrix} 0\\ \mathrm{a}_{J_{2x}}\\ 0\\ \mathrm{a}_{J_{2y}}\\ 0\\ \mathrm{a}_{J_{2z}}\end{bmatrix}}_{\mathbf{d}'(\mathbf{x})} \tag{4.32}$$

where $\mathbf{x} \triangleq [x\ \dot{x}\ y\ \dot{y}\ z\ \dot{z}]^T = [x_1\ x_2\ x_3\ x_4\ x_5\ x_6]^T$. For more details, reader can refer to Section 2.3.2. As per framework discussed in Section 4.1, the system dynamics Eq. (4.32) can be rewritten as

$$\underbrace{\begin{bmatrix} \dot{x}_1 \\ \dot{x}_2 \\ \dot{x}_3 \\ \dot{x}_4 \\ \dot{x}_5 \\ \dot{x}_6 \end{bmatrix}}_{\dot{\mathbf{x}}} = \underbrace{\begin{bmatrix} 0 & 1 & 0 & 0 & 0 & 0 \\ 3\omega^2 & 0 & 0 & 2\omega & 0 & 0 \\ 0 & 0 & 0 & 1 & 0 & 0 \\ 0 & -2\omega & 0 & 0 & 0 & 0 \\ 0 & 0 & 0 & 0 & 0 & 1 \\ 0 & 0 & 0 & 0 & -\omega^2 & 0 \end{bmatrix}}_{A} \underbrace{\begin{bmatrix} x_1 \\ x_2 \\ x_3 \\ x_4 \\ x_5 \\ x_6 \end{bmatrix}}_{\mathbf{x}} + \underbrace{\begin{bmatrix} 0 & 0 & 0 \\ 1 & 0 & 0 \\ 0 & 0 & 0 \\ 0 & 1 & 0 \\ 0 & 0 & 0 \\ 0 & 0 & 1 \end{bmatrix}}_{B} \mathbf{u} + \underbrace{\begin{bmatrix} d_1(\mathbf{x}) \\ d_2(\mathbf{x}) \\ d_3(\mathbf{x}) \\ d_4(\mathbf{x}) \\ d_5(\mathbf{x}) \\ d_6(\mathbf{x}) \end{bmatrix}}_{\mathbf{d}(\mathbf{x})} \tag{4.33}$$

where $[d_1(\mathbf{x})\ \ d_3(\mathbf{x})\ \ d_5(\mathbf{x})]^T = [0\ \ 0\ \ 0]^T$ and

$$d_2(\mathbf{x}) = 2\dot{\nu}_c x_4 + \ddot{\nu}_c x_3 + \dot{\nu}_c^2 x_1 - \frac{\mu}{\gamma}(x_1 + r_c) + \frac{\mu}{r_c^2} + \mathrm{a}_{\mathrm{J}_{2x}} - 3\omega^2 x_1 - 2\omega x_4 \tag{4.34}$$

$$d_4(\mathbf{x}) = -2\dot{\nu}_c x_2 - \ddot{\nu}_c x_1 + \dot{\nu}_c^2 x_3 - \frac{\mu}{\gamma} x_3 + \mathrm{a}_{\mathrm{J}_{2y}} + 2\omega x_2 \tag{4.35}$$

$$d_6(\mathbf{x}) = -\frac{\mu}{\gamma} x_5 + \mathrm{a}_{\mathrm{J}_{2z}} + \omega^2 x_5 \tag{4.36}$$

Note that even though the disturbance function components are deterministic, using these complex expressions in the necessary conditions of optimality is a challenge, especially in the costate equation. However, following the approach proposed here avoids this difficulty. As discussed in Section 4.1, the actual and observer dynamics are propagated using the numerical methods. At every step, based on the error information between the actual and observer state, the weight update rule Eq. (4.16) is evaluated. Without loss of generality, the basis function for approximating the disturbance is chosen as $\Phi_i(\mathbf{x}) = ((1/\eta) + (1/\eta)^2 + (1/\eta)^3 + (1/\eta)^4)\mathbf{x}$ for $i = 1, \ldots, 3$

where

$$\eta = \sqrt{(x^2 + y^2 + z^2)} \tag{4.37}$$

Next, using the updated weight $\hat{W}_i$ and the basis function $\Phi_i(\mathbf{x})$, the unknown function approximation $\hat{d}_i$ is calculated in each channel using Eq. (4.8) and subsequently used in the necessary conditions of optimality.

4.4 Results and Discussions

In this section, the significance of the Adaptive LQR theory-based guidance is shown by including the results for those cases for which the LQR philosophy fails, i.e., combination due to large relative distance and eccentricity of the chief satellite. Also an additional scenario is included to demonstrate the significance of the Adaptive LQR in presence of J_2 perturbation effect.

Simulation results provided in this section can be obtained from the program files provided in the folder named '*Adaptive LQR*'. For more details, refer to Section A.2.

4.4.1 Formation Flying in Elliptic Orbit and with Large Desired Relative Distance

First, the scenario in which the chief satellite is in an eccentric orbit and the desired relative distance is large is selected to demonstrate the capability of the Adaptive LQR philosophy. The initial condition and the desired final conditions for this simulation study are selected as per Table 2.5 in Chapter 2.

Figure 4.2 shows the trajectory plot of LQR and Adaptive LQR methods considering eccentricity $e = 0.05$ for chief satellite. From this figure, it can clearly be observed that the Adaptive LQR guidance performs much better than the LQR guidance in achieving the desired final condition since the neural network captures the nonlinearity and augments the nominal controller. Note that the position and velocity error at the final time for LQR is not zero as shown in Figs. 4.3 and 4.4, respectively, whereas these errors approach zero for the Adaptive LQR technique. Since the associated control history for these techniques look pretty similar, the difference between the two histories are plotted in Fig. 4.5. This small difference in the control trajectory of the adaptive controller leads to substantial enhancement of the performance.

4.4.2 Formation Flying in Elliptic Orbit and Small Desired Relative Distance with J_2 Perturbation

Next, the scenario considered corresponds to the eccentric orbit of the chief satellite in presence of the J_2 gravitational perturbation. The initial conditions considered here are given in Table 2.3 in Chapter 2.

Figure 4.6 shows the state trajectories with the application of the nominal LQR controller as well as with the Adaptive LQR controller. It can be seen that the trajectories with the application of the Adaptive LQR control solution achieve the desired final condition even in the presence of the J_2 perturbation. Note that the position and velocity error at the final time for LQR is not zero while those with the application of the Adaptive LQR controller approach to zero, as shown in Figs. 4.7 and 4.8, respectively. Figure 4.9 illustrates the performance of the neural network NN_2 for the dynamic variables. The solid line denotes the actual disturbance and the dotted line signifies the neural network approximation of the corresponding disturbance term. It can be seen that these networks successfully

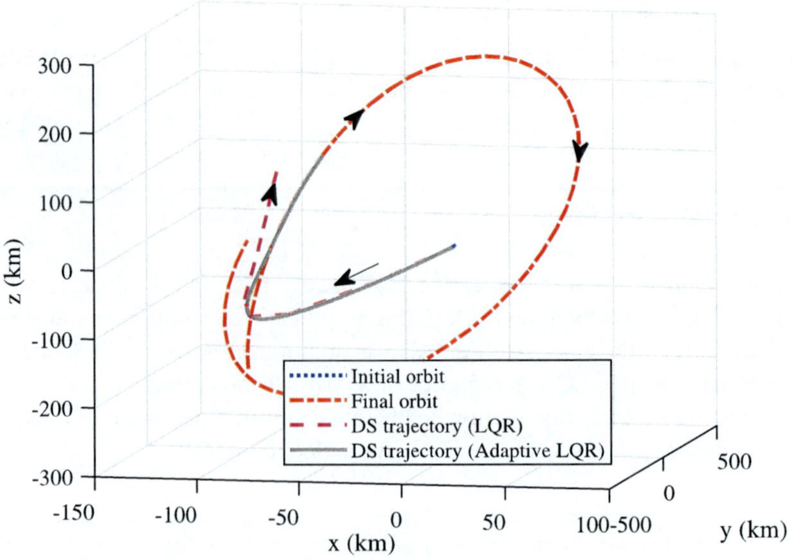

Fig. 4.2 Deputy satellite (DS) trajectory ($\rho_i = 1$ km, $\rho_f = 100$ km, $e = 0.05$)

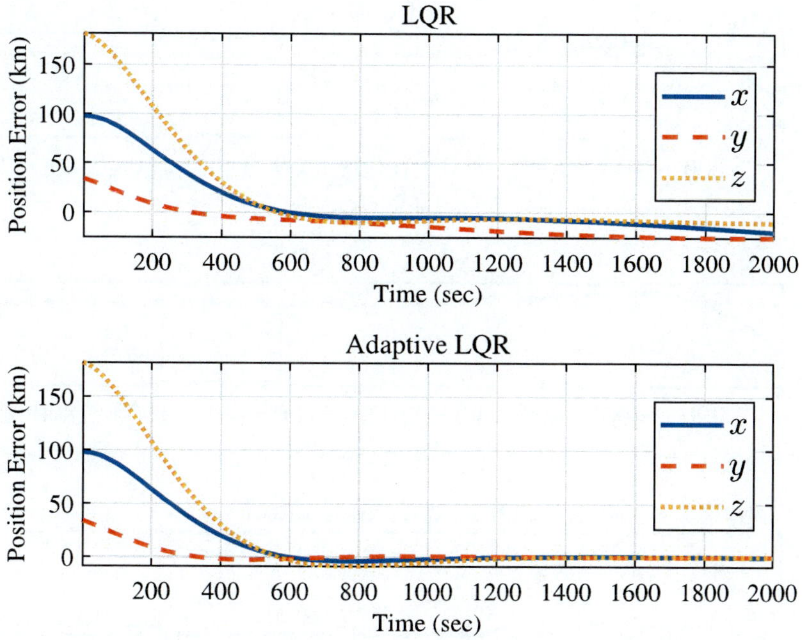

Fig. 4.3 Position error history ($\rho_i = 1$ km, $\rho_f = 100$ km, $e = 0.05$)

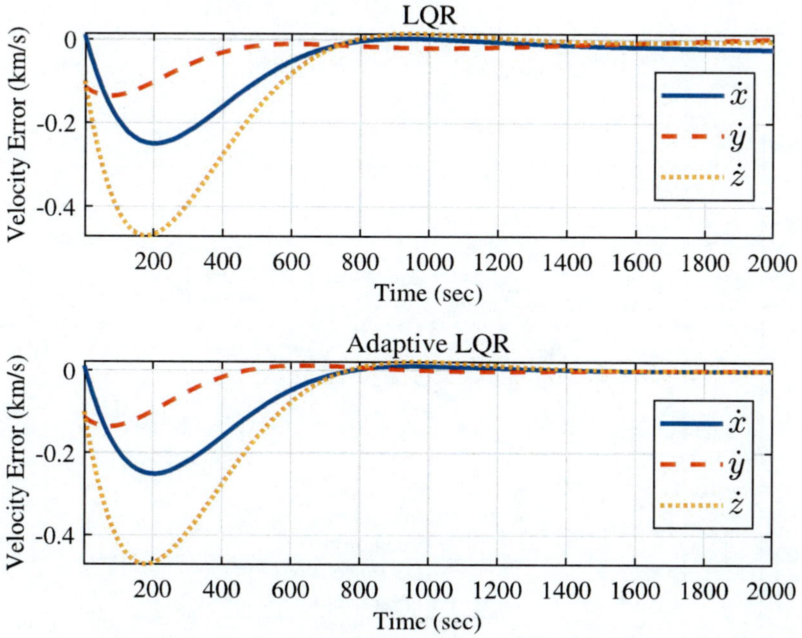

Fig. 4.4 Velocity error history ($\rho_i = 1$ km, $\rho_f = 100$ km, $e = 0.05$)

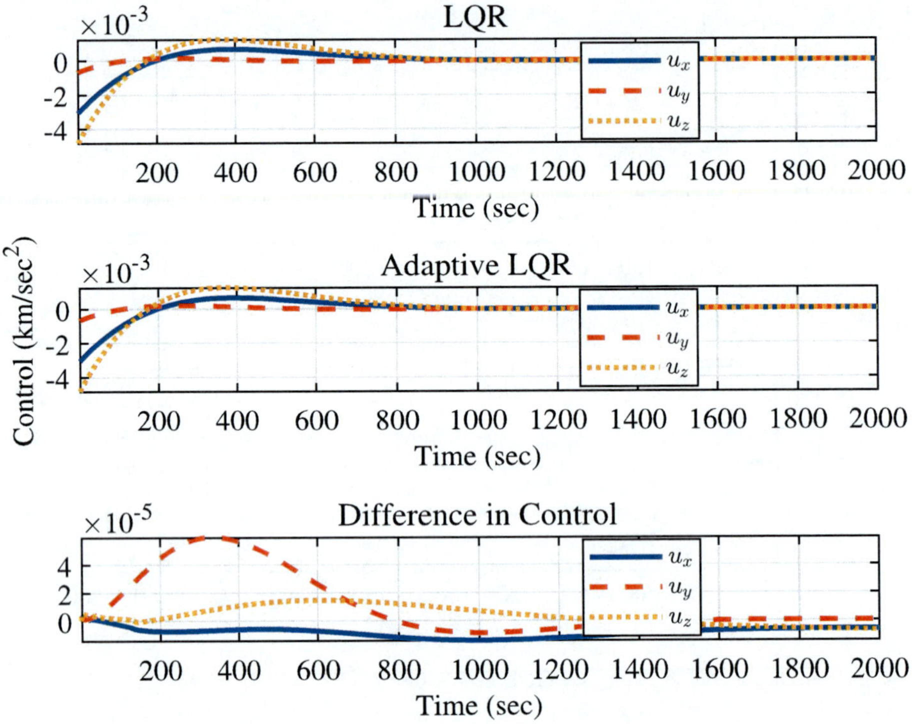

Fig. 4.5 Control history for DS ($\rho_i = 1$ km, $\rho_f = 100$ km, $e = 0.05$)

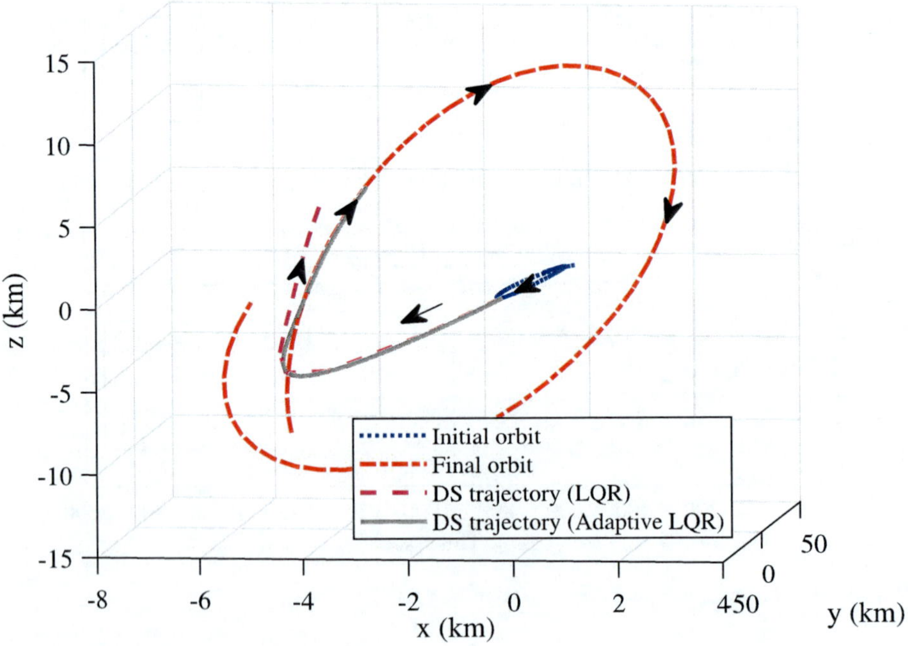

Fig. 4.6 Deputy satellite (DS) trajectory ($\rho_i = 1$ km, $\rho_f = 5$ km, $e = 0.05$)

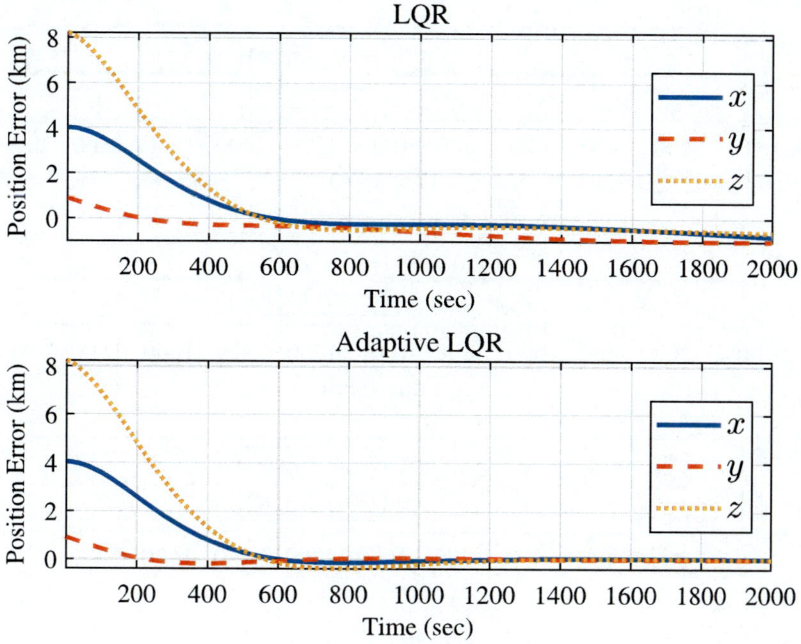

Fig. 4.7 Position error history ($\rho_i = 1\,\text{km}$, $\rho_f = 5\,\text{km}$, $e = 0.05$)

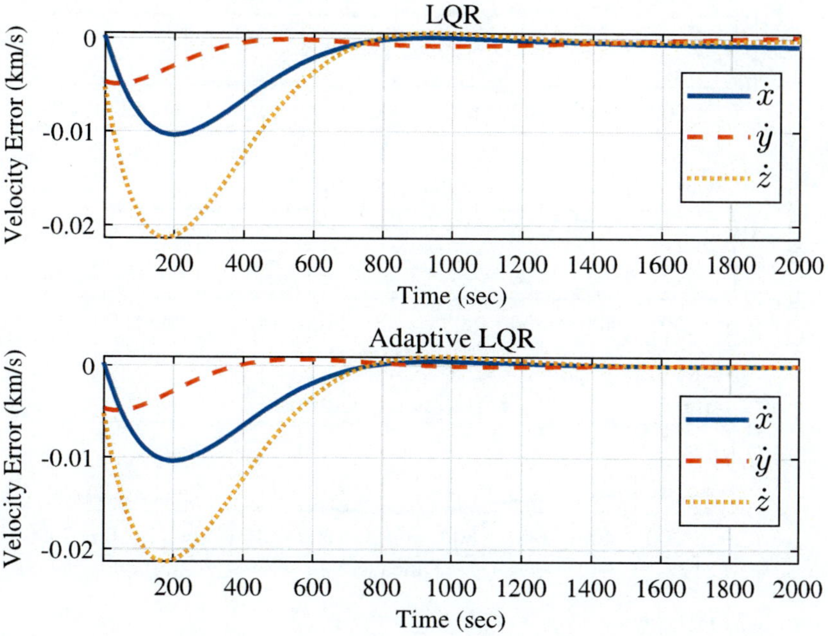

Fig. 4.8 Velocity error history ($\rho_i = 1\,\text{km}$, $\rho_f = 5\,\text{km}$, $e = 0.05$)

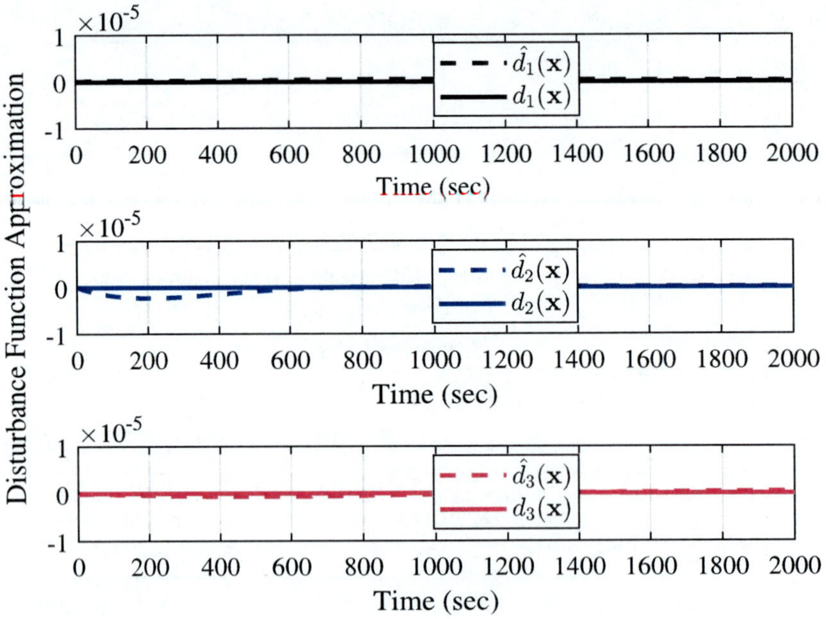

Fig. 4.9 Actual $\mathbf{d}(\mathbf{x})$ and approximated disturbances $\hat{\mathbf{d}}(\mathbf{x})$ for dynamic variables

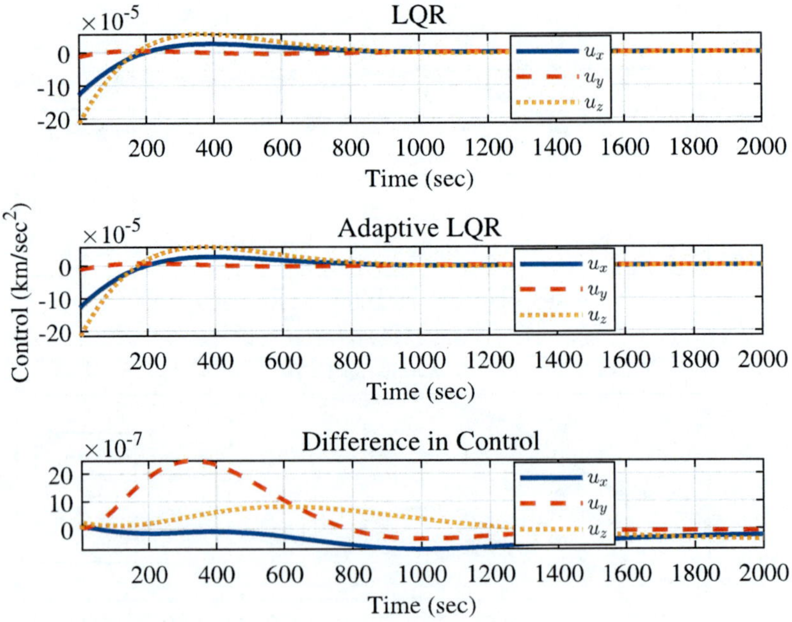

Fig. 4.10 Control history for DS ($\rho_i = 1$ km, $\rho_f = 5$ km, $e = 0.05$)

approximate the nonlinear function arising out of the plant approximation (to linear model) as well as exogenous J_2 disturbance. Figure 4.10 gives the associated control histories of LQR, Adaptive LQR as well as the difference between them. Even though the difference between the control histories are small, the Adaptive LQR leads to substantial enhancement of the performance.

4.5 Summary

This chapter concentrates on the synthesis of Adaptive LQR controller, where the LQR controller is modified online to compensate for the unmodeled dynamics and external perturbation. Satisfactory results are shown for spacecraft formation flying problem when the same is used as a guidance technique. It was shown that such an approach works in the presence of J_2 gravitational perturbation for cases where the chief satellite orbit is elliptic (with small eccentricity) and the desired relative distance is small. For more punishing cases, however, the reader is encouraged to see subsequent chapters.

References

1. Slotine, J.J., and W. Li. 1991. *Applied nonlinear control*. Prentice Hall.
2. Kim, B.S., and A.J. Calise. 1997. Nonlinear flight control using neural networks. *AIAA Journal of Guidance, Control and Dynamics* 20 (1): 26–33.
3. Ambati, P.R., and R. Padhi. 2017. Robust auto-landing of fixed-wing UAVs using neuro-adaptive design. *Control Engineering Practice* 60: 218–232.
4. Bryson, A.E., and Y.C. Ho. 1975. *Applied optimal control*. Hemisphere Publishing Corporation.
5. Naidu, D.S. 2002. *Optimal control systems*. CRC Press.
6. Lewis, F.L., D. Vrabie, and V.L. Syrmos. 2012. *Optimal control*. Wiley.

5 Adaptive Dynamic Inversion for Satellite Formation Flying

The benefit of satellite formation flying can truly be realized with greater mission flexibility such as higher inter-satellite separation, formation in elliptic orbits, etc. However, under the above-enhanced conditions, linear system dynamics based control design approaches fail to achieve the desired objectives. Even though the LQR philosophy inspired the SDRE approach discussed in Chapter 3 offers a limited solution, it suffers from the drawback that the success of the approach largely depends on the typical state-dependent coefficient form one adopts (which remains as 'art'). Moreover, if the eccentricity deviates significantly from circular orbit or separation distance requirement becomes large significantly, even SDRE can fail. The Adaptive LQR offers a fairly good solution to this issue, but introduces neural network learning concepts even for the system dynamics that is fairly known which can be handled directly. This brings in additional transients at the beginning of learning as well, which should preferably be avoided. In view of these observations, this chapter presents an alternate approach that need not be optimal, but can be successful under such realistic conditions as well.

The approach adopted in this chapter follows the widely popular nonlinear control design technique called *dynamic inversion (DI)* [1], which in turn is based on the philosophy of *feedback linearization* [2], followed by a control design approach that enforces a stable linear error dynamics. Even though the feedback linearization theory relies on differential geometric concepts and is quite vast, the dynamic inversion theory relies on the concept of input–output linearization coupled with the stability theory of linear systems. A rather straightforward engineering approach is presented here for quicker understanding.

It is also a fact that the dynamic inversion approach suffers from a severe drawback. As it relies heavily on manipulating the underlying nonlinear model, it becomes sensitive to modeling inaccuracies such as parametric inaccuracies, neglected dynamics, external disturbances, etc. A successful approach that is followed to address this issue is to augment it with an adaptive controller and a popular approach is the neuro-adaptive design [3, 4]. In this design, Lyapunov- based approach is used to train the neural network online, which ensures the stability of error dynamics and also bounds the neural network (NN) weights. The overall controller then becomes quite robust to the modeling inaccuracies and hence results in a powerful nonlinear controller.

In this chapter, brief summaries of the dynamic inversion and neuro-adaptive designs are outlined first in a generic sense for the benefit of the reader. The overall design technique is then successfully applied to the satellite formation flying problem. Simulation results include cases of formation

Electronic supplementary material The online version of this chapter (https://doi.org/10.1007/978-981-15-9631-5_5) contains supplementary material, which is available to authorized users.

S. Mathavaraj and R. Padhi, *Satellite Formation Flying*,
https://doi.org/10.1007/978-981-15-9631-5_5

flying with large eccentricities, large relative distance, and even considering J_2 perturbation, which are assumed to be unknown (thereby avoiding the associated complex modeling). Note that the same design also works equally well for the benign case of formation flying in circular orbit with small relative distance. Hence, to demonstrate its general utility, this benign case is presented first before presenting the more punishing case.

5.1 Dynamic Inversion: Generic Theory

Over the years, the dynamic inversion design technique has evolved as a good nonlinear control design technique [5]. As mentioned at the beginning of this chapter, this technique is essentially based on the philosophy of feedback linearization, wherein an appropriate coordinate transformation is carried out first so that the system appears to be a linear system in the transformed coordinates. This facilitates the usage of linear controller design techniques to design the controller in the transformed coordinates. Finally, the synthesized controller is transformed back to the original coordinate system to obtain a nonlinear state feedback controller. Subject to certain strong mathematical conditions [2], such an invertible nonlinear coordinate transformation turns out to be possible. It can also be mentioned here that even though any linear control design technique can be used in the transformed linear system dynamics, mostly the standard proportional–integral–derivative (PID) technique is preferred in practice to design the controller. Coupled with the inverse transformation, it essentially results in a meaningful nonlinear state feedback controller in the original coordinates [6, 7].

A brief outline of the dynamic inversion technique in the generic sense is presented below for completeness before applying it to the satellite formation flying problem. Also, embedding a PD form of the PID design, an alternate *easy-to-follow engineering approach* is presented here instead of going deep into mathematical details of coordinate transformation and associated algebra, thereby avoiding the associated complex mathematics completely.

Without loss of generality, for a better understanding of the underlying concept, let us focus on a class of nonlinear systems that are affine in control and represented by the following system dynamics:

$$\dot{\mathbf{x}} = f(\mathbf{x}) + g(\mathbf{x})\mathbf{u} \tag{5.1}$$

$$\mathbf{y} = h(\mathbf{x}) \tag{5.2}$$

where state $\mathbf{x} \in \Re^n$, control $\mathbf{u} \in \Re^m$, and performance output $\mathbf{y} \in \Re^p$. Let us further assume that $m = p$, i.e., the number of control variables is equal to the number of output variables (in the input–output sense, it is a square system). This, coupled with another assumption (see the comment after Eq. (5.5)), facilitates computation of the control command in closed form. The objective is to design a controller $\mathbf{u}(t)$ so that $\mathbf{y}(t) \to \mathbf{y}^*(t)$ as $t \to \infty$, where $\mathbf{y}^*(t)$ is the commanded reference signal.

To achieve this objective, first there is a need to compute the output dynamics. From Eq. (5.2), it is obvious that

$$\dot{\mathbf{y}} = \left[\frac{\partial h(\mathbf{x})}{\partial \mathbf{x}}\right]\dot{\mathbf{x}} \tag{5.3}$$

Next, using Eq. (5.1) in Eq. (5.3) the output dynamics can be written as

$$\dot{\mathbf{y}} = f_{\mathbf{y}}(\mathbf{x}) + g_{\mathbf{y}}(\mathbf{x})\mathbf{u} \tag{5.4}$$

where

$$f_{\mathbf{y}}(\mathbf{x}) \triangleq \left[\frac{\partial h(\mathbf{x})}{\partial \mathbf{x}}\right] f(\mathbf{x}), \quad g_{\mathbf{y}}(\mathbf{x}) \triangleq \left[\frac{\partial h(\mathbf{x})}{\partial \mathbf{x}}\right] B(\mathbf{x}) \tag{5.5}$$

Here, as another assumption, it is assumed that $g_{\mathbf{y}}(\mathbf{x})$ *non-singular* $\forall t$.

Next, defining $\boldsymbol{e} \triangleq (\mathbf{y} - \mathbf{y}^*)$, it is obvious that the objective can be achieved by synthesizing the controller $\mathbf{u}$ such that the following stable linear error dynamics is satisfied:

$$\dot{\boldsymbol{e}} + \mathbf{k}\,\boldsymbol{e} = 0 \tag{5.6}$$

where the gain matrix $\mathbf{k}$ is selected as a positive definite matrix. Note that if the error dynamics in Eq. (5.6) can be satisfied at all time, it will guarantee exponential stability of the error dynamics, thereby driving $\boldsymbol{e} \to 0$. Moreover, a usual and widely popular way to choose $\mathbf{k}$ is to select it as a diagonal matrix with positive entries in the diagonal elements.

Next, using the definition of $\boldsymbol{e}$, the Eq. (5.6) can be manipulated as follows:

$$(\dot{\mathbf{y}} - \dot{\mathbf{y}}^*) + \mathbf{k}(\mathbf{y} - \mathbf{y}^*) = 0 \tag{5.7}$$

$$f_{\mathbf{y}}(\mathbf{x}) + g_{\mathbf{y}}(\mathbf{x})\mathbf{u} - \dot{\mathbf{y}}^* + \mathbf{k}(\mathbf{y} - \mathbf{y}^*) = 0 \tag{5.8}$$

Equation (5.8) can be rewritten as

$$g_{\mathbf{y}}(\mathbf{x})\mathbf{u} = \boldsymbol{\beta} \tag{5.9}$$

where

$$\boldsymbol{\beta} \triangleq \left[-f_{\mathbf{y}}(\mathbf{x}) - \mathbf{k}(\mathbf{y} - \mathbf{y}^*) + \dot{\mathbf{y}}^*\right] \tag{5.10}$$

Since $g_{\mathbf{y}}(\mathbf{x})$ is assumed to be non-singular $\forall t$, from Eq. (5.10) the solution for the control variable can be written as

$$\mathbf{u} = \left[g_{\mathbf{y}}(\mathbf{x})\right]^{-1} \boldsymbol{\beta} \tag{5.11}$$

One can notice here that it is a fairly straightforward approach, leading to a *closed-form solution* of the control variable. Moreover, as long as the computed control from Eq. (5.11) is realized and implemented, it guarantees exponential stability of the output error dynamics. In the process, the technique serves as a universal gain scheduling controller, and hence, there is no need of extensive gain scheduling.

Even though this sounds very promising and lucrative, it should be kept in mind that this approach also suffers from a few important theoretical and practical issues. The very first issue arises from the assumption that $g_{\mathbf{y}}(\mathbf{x})$ is *non-singular* $\forall t$, which need not always be true. If this assumption gets violated for isolated small intervals of time, perhaps an easy engineering approach is not to update the control variable temporarily. Else, if it happens for large intervals of time, the formulation itself turns out to be erroneous. In that case, the usual recommendation is to select a different performance output $\mathbf{y}$ and the corresponding reference signal $\mathbf{y}^*$.

The second issue pops up because of the assumption that the same number of control inputs are available as the number of tracked outputs. However, this need not hold good in general. In case the number of control variables is lesser than the number of tracked outputs, no perfect tracking is possible [8]. However, if it is otherwise, i.e., the number of control variables is more than the number of tracked outputs, additional objectives are usually introduced in the problem objective to obtain a solution for the controller. One way of doing it leads to the concept of *optimal dynamic inversion* [9, 10], where not only perfect tracking is possible, but also it can be done with minimum control effort.

Another important issue arises from the fact that this design approach need not always depend on the first-order error dynamics. In fact, the exact order of the enforced error dynamics depends on the *relative degree* of the system and the output under consideration. The relative degree of each component of the output vector is defined as the number of times the variable needs to be differentiated so that the control variable appears explicitly. The total relative degree of the system, on the other hand, is defined as the summation of all individual relative degrees.

Without loss of generality, assuming that all components of the output vector are of relative degree 2, the following second-order error dynamics can be enforced to synthesize the controller:

$$\ddot{e} + \mathbf{k}_1 \dot{e} + \mathbf{k}_2 e = 0 \tag{5.12}$$

where the gain matrices $\mathbf{k}_1$, $\mathbf{k}_2$ need to be positive definite. Once again, the usual practice is to select these gain matrices as diagonal matrices with positive entries. Note that this also leads to the decoupling of output channels as a byproduct, which is another desirable feature. It can be mentioned here that the satellite formation flying problem satisfies this framework and Eq. (5.12) is used to derive the associated guidance command.

Yet another very important issue is the question of *stability of internal dynamics*, which essentially deals with the evolution of the untracked states with the application of the synthesized control. It essentially deals with the dynamics in the subspace not contained in the space spanned by the output space. One way of ensuring local stability of the internal dynamics for arbitrary reference commands is to ensure that the associated *zero dynamics* is locally stable. For more discussion on this issue, one can refer to [1, 2]. Details are not included here to contain the digression.

There is another potential issue in the dynamic inversion design. Because of modeling error and parameter inaccuracies, inversion of the model does not lead to the exact cancelation of the nonlinearities. Because of this, the technique essentially becomes sensitive to the parameter inaccuracies and unmodeled dynamics. Hence, there is a strong need to augment this technique with some adaptive and/or robust control design tools. A particular approach is based on the philosophy of *online model improvement*, where the system model in improved online by utilizing the error signal resulting out of the difference between the state vector projected from a disturbance observer and the actual state vector as sensed by the sensors [11] (it is assumed here that all states are measurable by the deployed sensors). The updated model is then utilized to resynthesize the controller. This results in an adaptive control architecture, which enhances the robustness of the closed-loop system against the unmodeled dynamics. In fact, this generic technique works with any nominal controller. However, whenever dynamic inversion is the baseline controller, the overall technique can be called as Adaptive dynamic inversion, the generic theory of which is discussed next for completeness.

5.2 Adaptive Dynamic Inversion: Generic Theory

In this approach, the key idea is to synthesize a set of linear-in-the-weight neural networks, which collectively capture the algebraic function that arises either because of the unmodeled (neglected) dynamics or because of the uncertainties in the parameters. A distinct characteristic of the adaptation procedure presented in this work is that it is independent of the technique used to design the nominal controller; and hence can be used in conjunction with any known control design technique. In this design, the controller synthesis is performed in a two-step design process for control realization. First, a disturbance observer is constructed and the actual states are enforced to follow the observer states. Simultaneously, however, the observer states are driven to the 'desired states', which are synthesized from a nominal (disturbance free) desired system dynamics. The entire process is described now with necessary details.

The focus here is on the class of nonlinear systems of Eqs. (5.13) and (5.14), which can be represented in the following structured form

$$\dot{\mathbf{x}}_1 = \mathbf{x}_2 \tag{5.13}$$

$$\begin{aligned}\dot{\mathbf{x}}_2 &= f(\mathbf{x}) + g(\mathbf{x})\,\mathbf{u} + d(\mathbf{x}) \\ &= f(\mathbf{x}_1, \mathbf{x}_2) + g(\mathbf{x}_1, \mathbf{x}_2)\,\mathbf{u} + \mathbf{d}(\mathbf{x}_1, \mathbf{x}_2)\end{aligned} \tag{5.14}$$

where, $\mathbf{x} = [\mathbf{x_1}^T\ \mathbf{x_2}^T]^T \in \Re^n$ represent the full state vector, whereas $\mathbf{x}_1 \in \Re^{n/2}$, $\mathbf{x}_2 \in \Re^{n/2}$ represent the kinematic and dynamic parts of the state vector, respectively. Another assumption here is that the system is point-wise controllable, which means the linearized system at every grid point of time is controllable in the sense of a linear time-invariant system [12]. The term $\mathbf{d}(\mathbf{x})$ represents the disturbance function, which is unknown and needs to be identified online. It is assumed that $\mathbf{d}(\mathbf{x})$ lies in a *compact set* so that it conforms to the universal function approximation property of neural networks [13–15]. It is also assumed that $\mathbf{d}(\mathbf{x})$ is smooth and slowly varying so that it can be captured by a neural network being trained online.

Next, assuming no uncertainty in the model, the dynamics of a 'desired nonlinear plant' can be represented as

$$\dot{\mathbf{x}}_{1_d} = \mathbf{x}_{2d} \tag{5.15}$$

$$\dot{\mathbf{x}}_{2_d} = f(\mathbf{x}_d) + g(\mathbf{x}_d)\,\mathbf{u}_d \tag{5.16}$$

where $\mathbf{x}_d = [\mathbf{x}_{1_d}^T\ \mathbf{x}_{2_d}^T]^T \in \Re^n$ represent the full desired state vector, $\mathbf{u}_d \in \Re^m$ represents the desired control.

The ultimate objective is to ensure that $[\mathbf{x_1}^T\ \mathbf{x_2}^T]^T \rightarrow [\mathbf{x_1}_d^T\ \mathbf{x_2}_d^T]^T$. However, since $\mathbf{d}(\mathbf{x})$ is unknown, it cannot be done in a straightforward manner. The key idea here is to first construct a 'disturbance observer', which is represented as

$$\dot{\mathbf{x}}_{1_a} = \mathbf{x}_2 + \mathbf{k}_{1a}(\mathbf{x}_1 - \mathbf{x}_{1a}) \tag{5.17}$$

$$\dot{\mathbf{x}}_{2_a} = f(\mathbf{x}) + g(\mathbf{x})\mathbf{u} + \hat{d}(\mathbf{x}) + \mathbf{k}_{2a}(\mathbf{x}_2 - \mathbf{x}_{2a}) \tag{5.18}$$

where $\mathbf{k}_{1a}, \mathbf{k}_{2a}$ are positive definite gain matrices, which is selected by the designer and tuned for best performance. The subscript 'a' is selected in Eqs. (5.17) and (5.18) to represent the approximate system dynamics. Here, $\mathbf{x}_a = [\mathbf{x}_{1_a}^T\ \mathbf{x}_{2_a}^T]^T \in \Re^n$ represent the full state vector of approximate system dynamics and $\hat{\mathbf{d}}(\mathbf{x})$ is the approximation to $\mathbf{d}(\mathbf{x})$.

After constructing this observer, the objective of $[\mathbf{x}_1{}^T\ \mathbf{x}_2{}^T]^T \rightarrow [\mathbf{x}_{1_d}{}^T\ \mathbf{x}_{2_d}{}^T]^T$ can be achieved by ensuring the following simultaneously.

(i) $\mathbf{x} \rightarrow \mathbf{x}_a$ as $t \rightarrow \infty$. This is accomplished by ensuring stability of the associated error dynamics using the *Lyapunov stability theory* [2], which, as a byproduct, results in training of the weights of the associated neural networks.

(ii) $\mathbf{x}_a \rightarrow \mathbf{x}_d$ as $t \rightarrow \infty$ which is accomplished through ensuring a stable linear error dynamics following the dynamic inversion philosophy, which, as a consequence leads to a closed-form expression of the control variable.

These two steps are discussed in detail next.

5.2.1 Ensuring $\mathbf{x} \rightarrow \mathbf{x}_a$: Design of a Disturbance Observer

The task here is to simultaneously ensure $\mathbf{x}_1 \rightarrow \mathbf{x}_{1_a}$, $\mathbf{x}_2 \rightarrow \mathbf{x}_{2_a}$ and $\hat{\mathbf{d}}(\mathbf{x}) \rightarrow \mathbf{d}(\mathbf{x})$ as the plant starts operating. Note that once this objective is met, the identified system dynamics through the observer dynamics can be a close representation of the actual system dynamics.

First $\mathbf{d}(\mathbf{x})$ can be written as $\mathbf{d}(\mathbf{x}) = \left[d_1(\mathbf{x}), \ldots, d_{n/2}(\mathbf{x})\right]^T$, where $d_i(\mathbf{x}),\ i = 1, \ldots, n/2$ is the ith component of $\mathbf{d}(\mathbf{x})$. Next, the idea is to approximate each $d_i(\mathbf{x})$ as $\hat{d}_i(\mathbf{x})$ in a linear-in-weight generic neural network representation, which can be written as

$$\hat{d}_i(\mathbf{x}) = \hat{W}_i^T \Phi_i(\mathbf{x}) \tag{5.19}$$

where $\hat{W}_i$ is the actual weight vector of the neural network and $\Phi_i(\mathbf{x})$ is the associated basis functions. The basis functions can either be chosen from a careful analysis of system dynamics for structured uncertainties (such as linearly appearing parameters) or selected as a set of generic functions (such as Gaussian functions) for unstructured uncertainties. Even though ideally $\hat{W}_i$ is a *constant* vector, it is assumed to be time-varying so that it can be updated with the evolution of time, leading to an online training process. However, it is found that all components of $\hat{W}_i$ evolve to steady-state (i.e., constant) values with the evolution of time.

From the *universal function approximation property* of neural networks [13–15], it is known that there exists an ideal neural network with an optimal weight W_i and basis function $\Phi_i(\mathbf{x})$ that approximates $\hat{d}_i(\mathbf{x})$ with accuracy ε_i. In other words, one can write

$$d_i(\mathbf{x}) = W_i^T \Phi_i(\mathbf{x}) + \varepsilon_i \tag{5.20}$$

The need for training the neural network arises because the ideal weight W_i is not known. In this development, it is assumed that both W_i and $\hat{W}_i \in \Re^{q_i \times 1}$.

Next, the task is to come up with a way of updating the weights of the neural networks (i.e., training the networks) such that the unknown function $d_i(\mathbf{x}), i = 1, \ldots, (n/2)$ can be captured. In this context, the error between actual and approximate kinematic and dynamic states in each channel is defined as

$$e_{1_i} \triangleq x_{1_i} - x_{1a_i} \tag{5.21}$$

$$e_{2_i} \triangleq x_{2_i} - x_{2a_i} \tag{5.22}$$

Taking the time derivative of Eqs. (5.21) and (5.22) and using the channel-wise information in Eqs. (5.13)–(5.18), one can write

$$\dot{e}_{1_i} = \dot{x}_{1_i} - \dot{x}_{1_{a_i}} = -k_{1a_i} e_{1_i} \tag{5.23}$$

$$\dot{e}_{2_i} = \dot{x}_{2_i} - \dot{x}_{2_{a_i}} = d_i(\mathbf{x}) - \hat{d}_i(\mathbf{x}) - k_{2a_i} e_{2_i} \tag{5.24}$$

Using the definition of $d_i(\mathbf{x})$ and $\hat{d}_i(\mathbf{x})$, Eqs. (5.23) and (5.24) can be rewritten as

$$\dot{e}_{1_i} = -k_{1a_i} e_{1_i} \tag{5.25}$$

$$\dot{e}_{2_i} = W_i^T \Phi_i(\mathbf{x}) + \varepsilon_i - \hat{W}_i^T \Phi_i(\mathbf{x}) - k_{2a_i} e_{2_i} \tag{5.26}$$

Defining the error in weights of the ith network as $\tilde{W}_i \triangleq W_i - \hat{W}_i$, one can observe that $\dot{\tilde{W}}_i = -\dot{\hat{W}}_i$ as W_i is a constant weight vector. Next, the aim is to derive a weight update rule such that the weights of the approximating networks $\hat{W}_i$ approach the ideal weights W_i.

Next, a positive definite Lyapunov function candidate is considered as

$$V = \sum_{i=1}^{n/2} (V_{1_i} + V_{2_i}) \tag{5.27}$$

where

$$V_{1_i}(e_{1_i}) = \frac{e_{1_i}^2}{2} \tag{5.28}$$

$$V_{2_i}(e_{2_i}, \tilde{W}_i) = \beta_i \frac{e_{2_i}^2}{2} + \frac{\tilde{W}_i^T \tilde{W}_i}{2\gamma_i} + \left[\frac{\partial d_i(\mathbf{x})}{\partial \mathbf{x}} - \frac{\partial \hat{d}_i(\mathbf{x})}{\partial \mathbf{x}}\right]^T \frac{\Theta_i}{2} \left[\frac{\partial d_i(\mathbf{x})}{\partial \mathbf{x}} - \frac{\partial \hat{d}_i(\mathbf{x})}{\partial \mathbf{x}}\right] \tag{5.29}$$

where β_i and γ_i are selected as positive constants and Θ_i is selected positive definite matrix. One can observe that the Lyapunov function V_{2_i} has the following three components:

(i) e_{2_i}, the error in states of ith channel,
(ii) $\tilde{W}_i$, the error in weights of ith network, and
(iii) $\left[\frac{\partial d_i(\mathbf{x})}{\partial \mathbf{x}} - \frac{\partial \hat{d}_i(\mathbf{x})}{\partial \mathbf{x}}\right]$, the error in partial derivative of ith unknown function.

It can be mentioned here that the third (partial derivative) term in the Lyapunov function V_{2_i} leads to a *directional learning* of the disturbance function. It can be noted that guaranteeing the stability of the error dynamics Eq. (5.26) through V_{2_i} for all $i = 1, \ldots, (n/2)$ ensures not only capturing of the disturbance function $\mathbf{d}(\mathbf{x})$ through capturing the weights of the neural networks but also the gradient of the disturbance functions in each channel. This leads to minimization of transient effects. Note that ensuring asymptotic stability is not possible unless the disturbance function can be captured as a 'structured uncertainty'. In general, one has to remain satisfied with 'practical stability' with small error bounds, details of which will be clear from the following discussion.

Using Eqs. (5.19) and (5.20), one can rewrite Eq. (5.29) as

$$\begin{aligned} V_{2_i}(e_{2_i}, \tilde{W}_i) &= \beta_i \frac{e_{2_i}^2}{2} + \frac{\tilde{W}_i^T \tilde{W}_i}{2\gamma_i} + (W_i - \hat{W}_i)^T \left[\frac{\partial \Phi_i}{\partial \mathbf{x}}\right] \frac{\Theta_i}{2} \left[\frac{\partial \Phi_i}{\partial \mathbf{x}}\right]^T (W_i - \hat{W}_i) \\ &= \beta_i \frac{e_{2_i}^2}{2} + \frac{\tilde{W}_i^T \tilde{W}_i}{2\gamma_i} + \tilde{W}_i^T \left[\frac{\partial \Phi_i}{\partial \mathbf{x}}\right] \frac{\Theta_i}{2} \left[\frac{\partial \Phi_i}{\partial \mathbf{x}}\right]^T \tilde{W}_i \end{aligned} \tag{5.30}$$

Next, the time derivative of the Lyapunov function Eq. (5.27) is written as follows:

$$\dot{V} = \sum_{i=1}^{n/2} (\dot{V}_{1_i} + \dot{V}_{2_i}) \tag{5.31}$$

where

$$\dot{V}_{1_i} = e_{1_i} \dot{e}_{1_i} \tag{5.32}$$

$$\dot{V}_{2_i} = \beta_i e_{2_i} \dot{e}_{2_i} - \frac{\tilde{W}_i^T \dot{\hat{W}}_i}{\gamma_i} - \tilde{W}_i^T \left[\frac{\partial \Phi_i}{\partial \mathbf{x}}\right] \Theta_i \left[\frac{\partial \Phi_i}{\partial \mathbf{x}}\right]^T \dot{\hat{W}}_i \tag{5.33}$$

where the higher order partial derivative terms are assumed to be zero. Substituting the error dynamics of e_{1_i} (refer to Eq. (5.25)) in Eq. (5.32) results in

$$\dot{V}_{1_i} = -k_{1a_i} e_{1_i}^2 \tag{5.34}$$

Next, substituting the error dynamics Eq. (5.26) in Eq. (5.33) results in

$$\dot{V}_{2_i} = \beta_i e_{2_i} \tilde{W}_i^T \Phi_i(\mathbf{x}) + \beta_i e_{2_i} \varepsilon_i - \beta_i k_{2a_i} e_{2_i}^2 - \frac{\tilde{W}_i^T \dot{\hat{W}}_i}{\gamma_i} - \tilde{W}_i^T \left[\frac{\partial \Phi_i}{\partial \mathbf{x}}\right] \Theta_i \left[\frac{\partial \Phi_i}{\partial \mathbf{x}}\right]^T \dot{\hat{W}}_i \tag{5.35}$$

Note that the objective in this exercise is to come up with a meaningful condition that will assure $\dot{V}_{2_i}$ is negative, so that the Lyapunov stability theory can be used. However, one can notice here that $\tilde{W}_i$ is unknown and hence nothing can be said about $\dot{V}_{2_i}$. To proceed with the analysis avoiding this difficulty is carried out by collecting and equating the coefficient of $\tilde{W}_i$ to zero.

$$\left(\frac{I_{q_i}}{\gamma_i} + \left[\frac{\partial \Phi_i}{\partial \mathbf{x}}\right] \Theta_i \left[\frac{\partial \Phi_i}{\partial \mathbf{x}}\right]^T\right) \dot{\hat{W}}_i - \beta_i e_{2_i} \Phi_i(\mathbf{x}) = 0 \tag{5.36}$$

where I_{q_i} is the identity matrix of dimension $q_i \times q_i$. However, quite fortunately, this also results in following weight update rule for the ith neural network in continuous time.

$$\dot{\hat{W}}_i = \beta_i e_{2_i} \left(\frac{I_{q_i}}{\gamma_i} + \left[\frac{\partial \Phi_i}{\partial \mathbf{x}}\right] \Theta_i \left[\frac{\partial \Phi_i}{\partial \mathbf{x}}\right]^T\right)^{-1} \Phi_i(\mathbf{x}) \tag{5.37}$$

The left over terms from Lyapunov derivative $\dot{V}_{2_i}$ in Eq. (5.35) gives us

$$\dot{V}_{2_i} = \beta_i e_{2_i} \varepsilon_i - k_{2a_i} \beta_i e_{2_i}^2 \tag{5.38}$$

for $\dot{V}_{2_i} < 0$ leads to a condition

$$|e_i| > \frac{|\varepsilon_i|}{k_{2a_i}} \tag{5.39}$$

It has to be noted that $\dot{V}_{1_i}$ for all t is negative. Hence all the components of $\dot{V}$ are negative, which in turn lead to stable error dynamics. Therefore, if the network weights are updated based on the rule given in Eq. (5.37), then the identification happens as long as absolute error is greater than certain value as in Eq. (5.39). It has to be noted that by increasing k_{2a_i}, error bound can be theoretically reduced.

The philosophy then is to use the observer dynamics to synthesize the controller. After the transient process is over, the dynamics becomes a close representation of the actual dynamics. However, it is obvious that once the observer dynamics is used to synthesize the controller, it essentially results in an adaptive control structure. In general, this adaptive controller turns out to be a very good robust controller for modeling inaccuracies. This is because the combined model, which is a combination of the offline gray-box model and online black-box model, becomes a very close representation of the actual plant, thereby reducing the modeling inaccuracy substantially. However, it is worth emphasizing again that as and when any improvement takes place in the model after extensive testing on ground, the same adaptive philosophy can be utilized almost immediately within the same controller structure. Moreover, it also has a disturbance rejection property because of the observer in the loop. Note that, since it is essentially a model improvement tool, it can be used in conjunction with any nonlinear control design technique as well.

5.2.2 Ensuring $\mathbf{x}_a \rightarrow \mathbf{x}_d$: Control Synthesis

In the context of dynamic inversion controller (refer to Section 5.1), $\mathbf{x}_a \rightarrow \mathbf{x}_d$ can be ensured through following second-order error dynamics:

$$(\ddot{\mathbf{x}}_{1_a} - \ddot{\mathbf{x}}_{1_d}) + \mathbf{k}_1(\dot{\mathbf{x}}_{1_a} - \dot{\mathbf{x}}_{1_d}) + \mathbf{k}_2(\mathbf{x}_{1_a} - \mathbf{x}_{1_d}) = 0 \tag{5.40}$$

where $\mathbf{k}_1$ and $\mathbf{k}_2$ are positive definite gains matrices, $\ddot{\mathbf{x}}_{1_a} = \dot{\mathbf{x}}_{2_a}, \ddot{\mathbf{x}}_{1_d} = \dot{\mathbf{x}}_{2_d}$. Substituting the approximate plant model Eq. (5.18) into Eq. (5.40) results in

$$\begin{aligned} f(\mathbf{x}) + g(\mathbf{x})\mathbf{u} + \hat{d}(\mathbf{x}) + \mathbf{k}_{2a}(\mathbf{x}_2 - \mathbf{x}_{2a}) - \ddot{\mathbf{x}}_{1_d} \\ + \mathbf{k}_1(\dot{\mathbf{x}}_{1_a} - \dot{\mathbf{x}}_{1_d}) + \mathbf{k}_2(\mathbf{x}_{1_a} - \mathbf{x}_{1_d}) = 0 \end{aligned} \tag{5.41}$$

Simplifying Eq. (5.41) for $\mathbf{u}$, the following equation is obtained:

$$\begin{aligned} \mathbf{u} = [g(\mathbf{x})]^{-1}[-f(\mathbf{x}) - \hat{d}(\mathbf{x}) - \mathbf{k}_{2a}(\mathbf{x}_2 - \mathbf{x}_{2a}) + \ddot{\mathbf{x}}_{1_d} \\ - \mathbf{k}_1(\dot{\mathbf{x}}_{1_a} - \dot{\mathbf{x}}_{1_d}) - \mathbf{k}_2(\mathbf{x}_{1_a} - \mathbf{x}_{1_d})] \end{aligned} \tag{5.42}$$

This closed-form expression for control $\mathbf{u}$ is valid provided $[g(\mathbf{x})]^{-1}$ exists for all values $\mathbf{x}$. It is to be noted that for the control synthesis, the unknown function $\hat{d}(\mathbf{x})$ is available through neural network approximation, as explained in Section 5.2.1.

5.3 SFF Problem Formulation in Di Framework

The kinematic and dynamic state variables (as per the framework discussed in Section 5.2) for satellite formation flying problem is defined as follows:

$$\mathbf{x}_1 \triangleq [x\ y\ z]^T = [x_1\ x_3\ x_5]^T \tag{5.43}$$

$$\mathbf{x}_2 \triangleq [\dot{x}\ \dot{y}\ \dot{z}]^T = [x_2\ x_4\ x_6]^T \tag{5.44}$$

where $\mathbf{x}_1$ and $\mathbf{x}_2$ represents the relative position and velocity of the deputy satellite in Hill's frame respectively. Now let us consider the actual plant model that includes the J_2 perturbation (refer to Section 2.3.2). In this case, the system dynamics is written as

$$\underbrace{\begin{bmatrix} \dot{x}_1 \\ \dot{x}_3 \\ \dot{x}_5 \end{bmatrix}}_{\dot{\mathbf{x}}_1} = \underbrace{\begin{bmatrix} x_2 \\ x_4 \\ x_6 \end{bmatrix}}_{\mathbf{x}_2} \tag{5.45}$$

$$\underbrace{\begin{bmatrix} \dot{x}_2 \\ \dot{x}_4 \\ \dot{x}_6 \end{bmatrix}}_{\dot{\mathbf{x}}_2} = \underbrace{\begin{bmatrix} 2\dot{\nu}_c x_4 + \ddot{\nu}_c x_3 + \dot{\nu}_c^2 x_1 - \frac{\mu}{\gamma}(x_1 + r_c) + \frac{\mu}{r_c^2} \\ -2\dot{\nu}_c x_2 - \ddot{\nu}_c x_1 + \dot{\nu}_c^2 x_3 - \frac{\mu}{\gamma} x_3 \\ -\frac{\mu}{\gamma} x_5 \end{bmatrix}}_{f(\mathbf{x})} + \underbrace{\begin{bmatrix} 1\ 0\ 0 \\ 0\ 1\ 0 \\ 0\ 0\ 1 \end{bmatrix}}_{g(\mathbf{x})} \mathbf{u} + \underbrace{\begin{bmatrix} a_{J_{2x}} \\ a_{J_{2y}} \\ a_{J_{2z}} \end{bmatrix}}_{d(\mathbf{x})} \tag{5.46}$$

where $f(\mathbf{x})$, $g(\mathbf{x})$, $d(\mathbf{x})$ are defined as in Eq. (5.46). It can be noted that the difference between the actual and the nominal model is the J_2 perturbation which is assumed to be unknown disturbance $d(\mathbf{x})$ for the analysis. Without loss of generality, the basis function for approximating the disturbance is chosen as $\Phi_i(\mathbf{x}) = ((1/\eta) + (1/\eta)^2 + (1/\eta)^3 + (1/\eta)^4)\mathbf{x}$ for $i = 1, \ldots, 3$
where

$$\eta = \sqrt{(x^2 + y^2 + z^2)} \tag{5.47}$$

Next, the actual and approximate dynamics (as discussed in Section 5.2.1) are propagated using the numerical methods. At every step, based on the error information between the actual and approximate state, the weight update rule Eq. (5.37) is evaluated by utilizing the selected basis function $\Phi_i(\mathbf{x})$. Next, using the updated weight $\hat{W}_i$ and the basis function, the unknown function approximation $\hat{d}_i$ is calculated in each channel using Eq. (5.19). Subsequently, the objective is to evaluate the adaptive control based on dynamic inversion philosophy discussed in Section 5.2.2. It is to be noted that for the satellite formation flying problem $g(\mathbf{x})$ is a constant and square matrix, which is invertable $\forall t$.

5.4 Results and Discussions

In this section, the significance of the adaptive control based on dynamic inversion theory-based guidance is shown by including the results for formation flying in high eccentricity and large desired relative distance scenario. However, simple formation flying problems with circular orbit and small desired relative distance is presented first to demonstrate its generality.

Simulation results provided in this section can be obtained from the program files provided in the folder named '*Adaptive DI*'. For more details, refer to Section A.3.

5.4.1 Formation Flying in Circular Orbit and with Small Desired Relative Distance

The initial condition and the desired final conditions for this simulation study are selected as per Table 2.2 in Chapter 2.

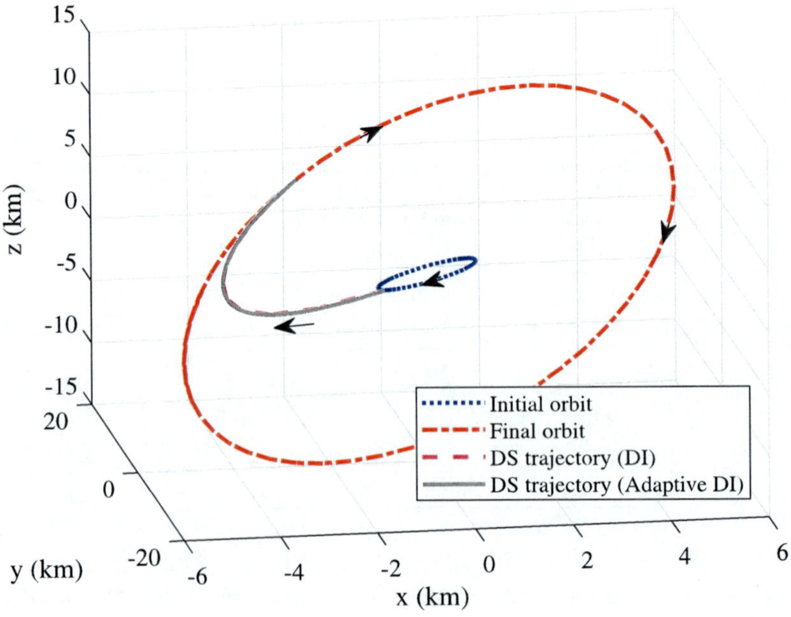

Fig. 5.1 Deputy satellite (DS) trajectory ($\rho_i = 1\,\text{km}$, $\rho_f = 5\,\text{km}$, $e = 0.0$)

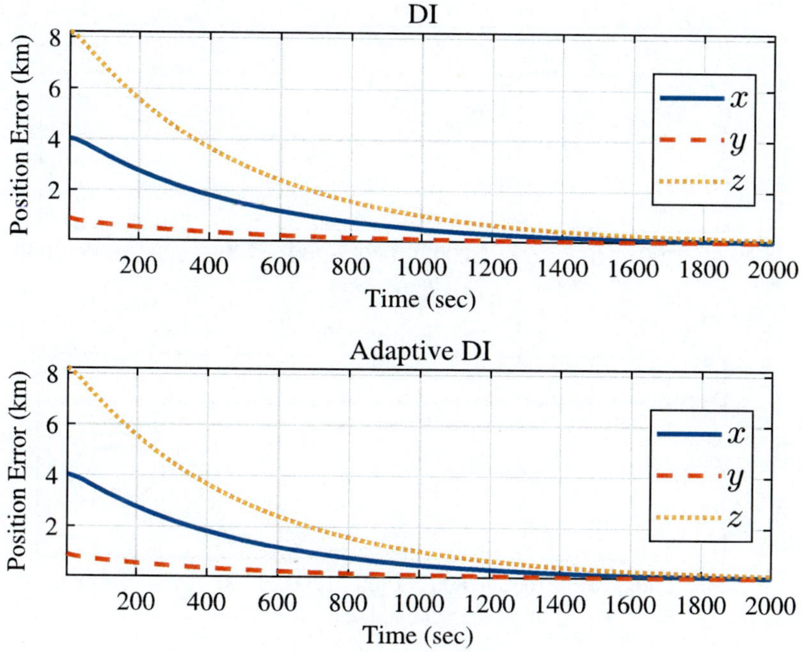

Fig. 5.2 Position error history ($\rho_i = 1$ km, $\rho_f = 5$ km, $e = 0.0$)

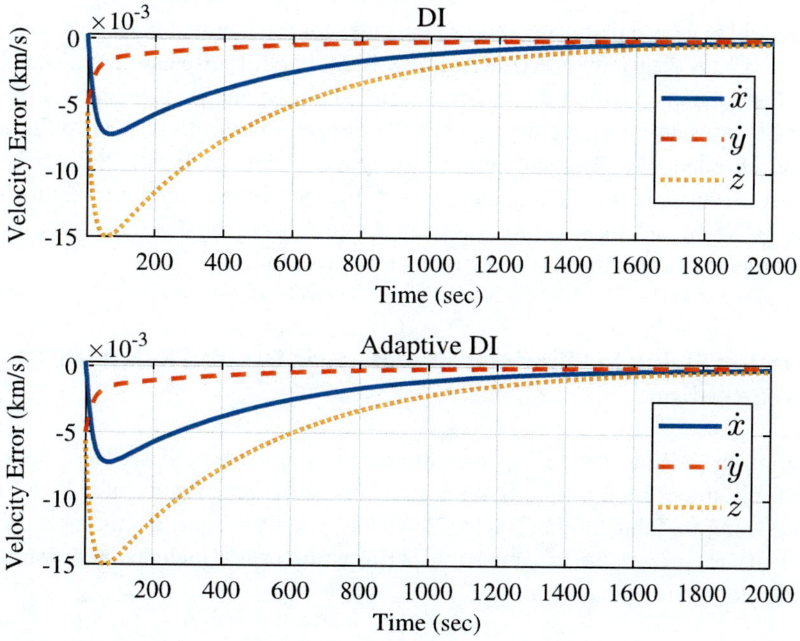

Fig. 5.3 Velocity error history ($\rho_i = 1$ km, $\rho_f = 5$ km, $e = 0.0$)

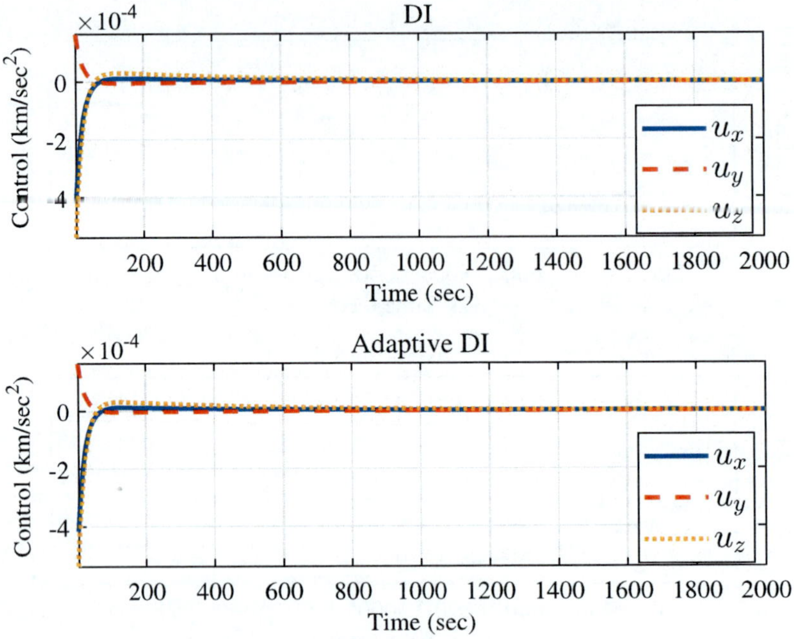

Fig. 5.4 Control history for deputy satellite ($\rho_i = 1$ km, $\rho_f = 5$ km, $e = 0.0$)

The SFF problem is solved first using the DI controller for the nominal plant with a circular orbit of the chief satellite (i.e., eccentricity $e = 0$) with a known radius. It is also assumed that no disturbance, such as the J_2 disturbance acts on the system. Figure 5.1 shows the performance of the nominal DI controller. The deputy satellite trajectory reaches the desired orbit with respect to the chief satellite using a nominal DI controller. The state error history are plotted in Figs. 5.2 and 5.3, which show how the state errors reduce along the trajectory to achieve the desired condition. Figure 5.4 shows the control solution of DI and adaptive control based on DI to achieve the desired orbit. Since there is no disturbance in the plant model, there is no difference between the two.

5.4.2 Formation Flying in Elliptic Orbit and Large Desired Relative Distance with J_2 Perturbation

In this simulation, the scenario is shifted to a more practical extreme case. Here the chief satellite is assumed to fly in the elliptic orbit having a relatively large eccentricity of 0.2 and large desired relative distance (refer to Table 2.6 in Chapter 2). The system dynamics, in this case, is considered as the 'nominal plant'. Moreover, the exogenous J_2 perturbation is also included in the simulation study, which is assumed to be the 'actual plant'.

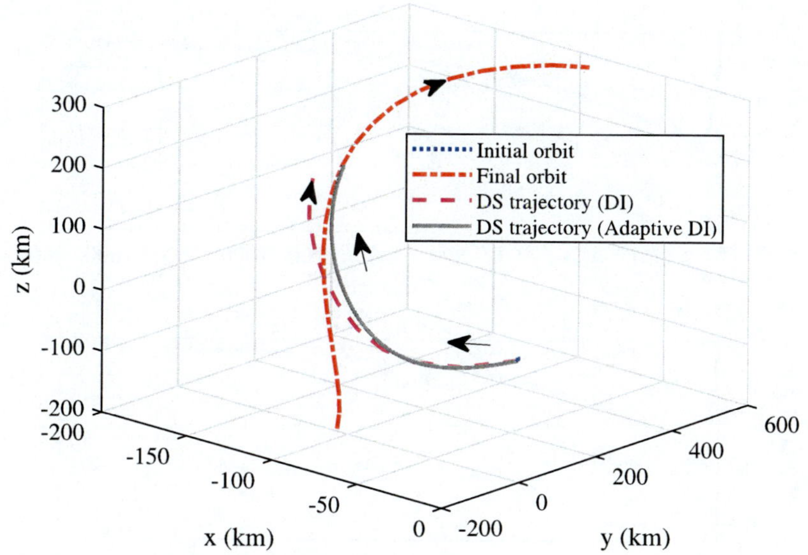

Fig. 5.5 Deputy satellite (DS) trajectory ($\rho_i = 1\,\text{km}$, $\rho_f = 100\,\text{km}$, $e = 0.2$)

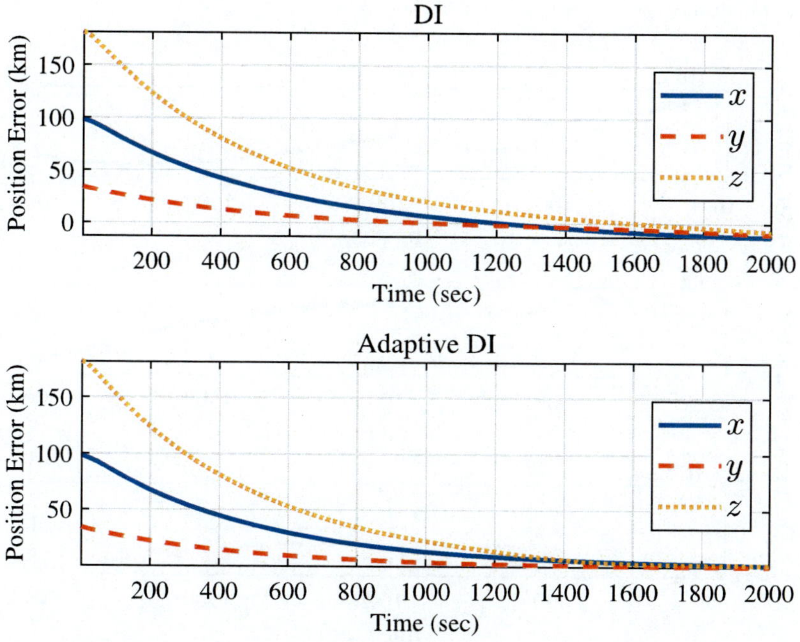

Fig. 5.6 Position error history ($\rho_i = 1\,\text{km}$, $\rho_f = 100\,\text{km}$, $e = 0.2$)

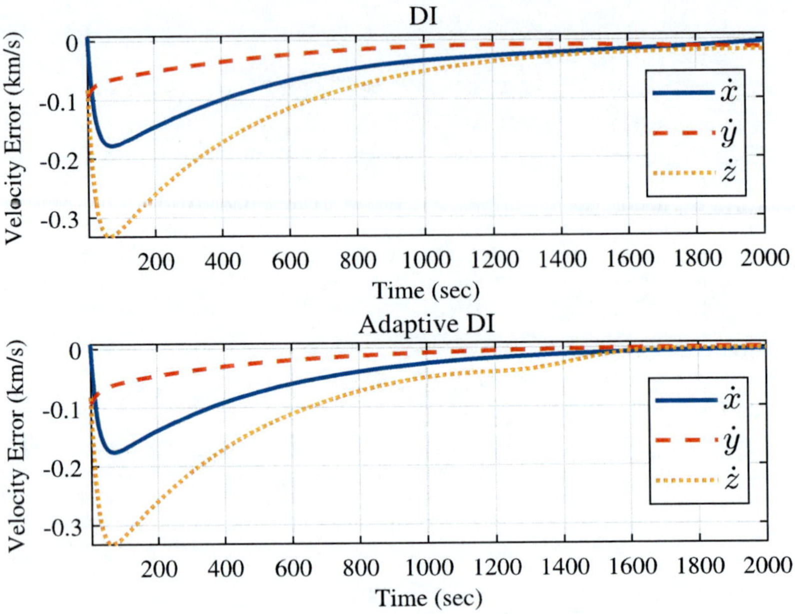

Fig. 5.7 Velocity error history ($\rho_i = 1$ km, $\rho_f = 100$ km, $e = 0.2$)

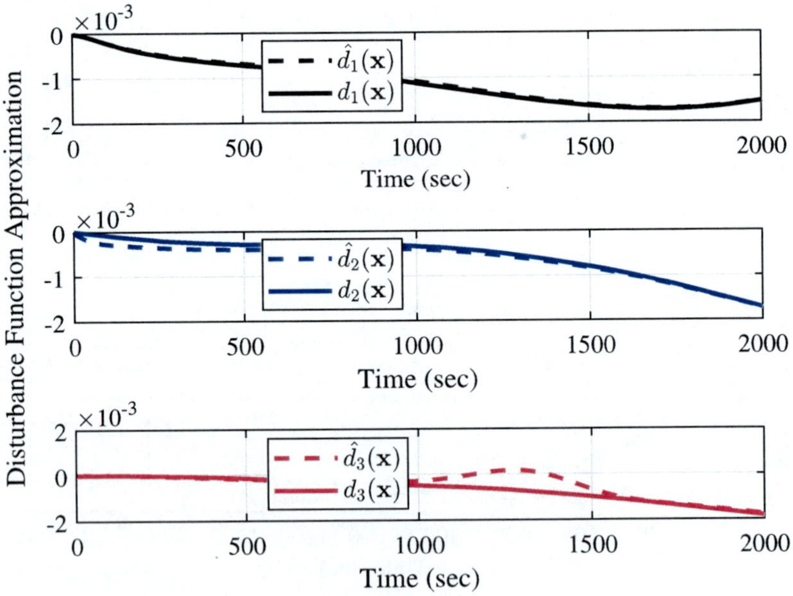

Fig. 5.8 Actual $\boldsymbol{d}(\mathbf{x})$ and approximated disturbances $\hat{\boldsymbol{d}}(\mathbf{x})$ for dynamic variables

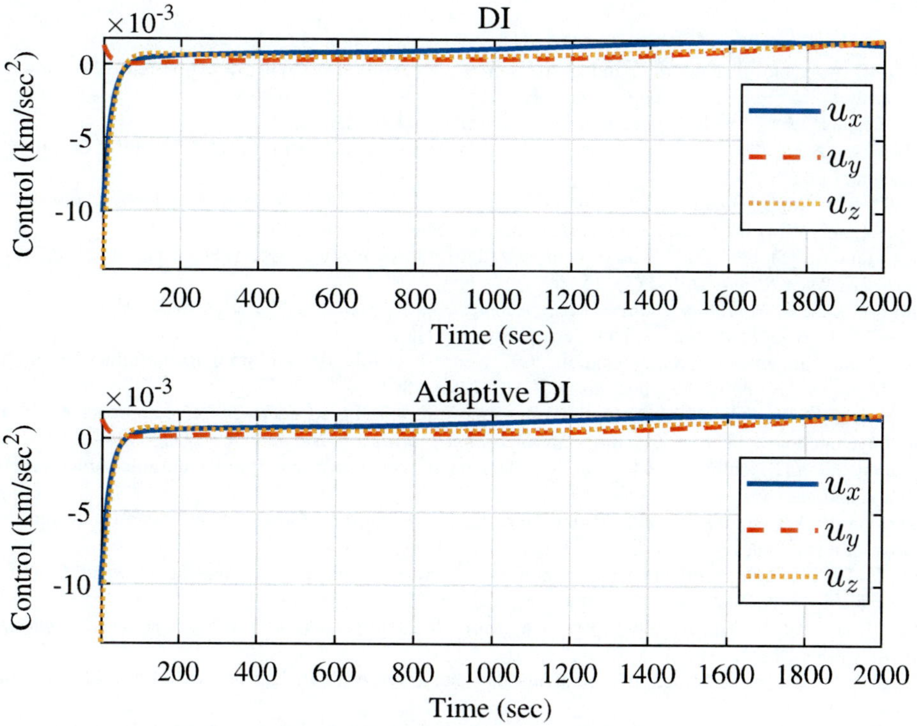

Fig. 5.9 Control history for deputy satellite ($\rho_i = 1$ km, $\rho_f = 100$ km, $e = 0.2$)

Figure 5.5 shows the trajectory of the actual plant (with disturbance) with adaptive nonlinear controller and trajectory of actual plant with a nominal controller. It can be inferred that the trajectory with the application of the adaptive nonlinear controller achieves the desired final condition, whereas the nominal trajectory (i.e., with the application of only the dynamic inversion controller) diverges. Hence, it is obvious that the suggested adaptive controller increases the robustness of the closed-loop system in presence of unknown disturbance.

Note that the position and velocity error at the final time for DI is not zero while adaptive control based on DI approaches zero as shown in Figs. 5.6 and 5.7, respectively. Figure 5.8 illustrates the satisfactory performance of three neural networks which is used to approximate the exogenous J_2 disturbance. The solid line denotes the actual disturbance, and the dotted line signifies the neural network approximation of the corresponding disturbance term. Figure 5.9 gives the details of the associated control histories.

5.5 Summary

In this chapter, we presented an online adaption scheme for satellite formation flying, which synthesizes an adaptive control to compensate for the unknown disturbances using dynamic inversion as baseline controller. From the results, it is found out that the neuro-adaptive dynamic controller is superior in terms of disturbance capturing when compared to the dynamic inversion-based nominal controller. We emphasize here that implementing online neural network adaptive DI controller is rather easy and straightforward and it also guarantees the mission objective of minimum terminal error, even in the presence of the external J_2 perturbation in the model.

References

1. Enns, D., D. Bugajski, R. Hendrick, and G. Stein. 1994. Dynamic inversion: An evolving methodology for flight control design. *International Journal of control* 59 (1): 71–91.
2. Slotine, J.J., and W. Li. 1991. *Applied nonlinear control*. Prentice Hall.
3. Kim, B.S., and A.J. Calise. 1997. Nonlinear flight control using neural networks. *AIAA Journal of Guidance, Control and Dynamics* 20 (1): 26–33.
4. Padhi, R., S.N. Balakrishnan, and N. Unnikrishnan. 2007b. Model following neuro-adaptive control design for non-square, non-affine nonlinear systems. *IET Control Theory Application* 1 (6): 1650–1661.
5. Wang, Q., and R.F. Stengel. 2004. Robust nonlinear flight control of a high-performance aircraft. *IEEE Transactions on Control Systems Technology* 13 (1): 15–26.
6. Marquez, H.J. 2003. *Nonlinear control systems: Analysis and design*, vol. 161, John Wiley Hoboken.
7. Kahlil, H. 2002. *Nonlinear systems*. New Jersey: Prentice Hall.
8. Li, Y., N. Sundararajan, and P. Saratchandran. 2001. Neuro-controller design for nonlinear fighter aircraft maneuver using fully tuned RBF networks. *Automatica* 37 (8): 1293–1301.
9. Padhi, R., and M. Kothari. 2007. An optimal dynamic inversion-based neuro-adaptive approach for treatment of chronic myelogenous leukemia. *Computer Methods and Programs in Biomedicine* 87 (3): 208–224.
10. Mathavaraj, S., and R. Padhi. 2019. Optimally allocated nonlinear robust control of a reusable launch vehicle during re-entry. *Unmanned Systems*.
11. Rajasekaran, J., A. Chunodkar, and R. Padhi. 2009. Structured model-following neuro-adaptive design for attitude maneuver of rigid bodies. *Control Engineering Practice* 17 (6): 676–689.
12. Cloutier, J. 1997. State-dependent Riccati equation techniques: An overview. In *American Control Conference*, vol. 2, 932–936. Air Force Armament Directorate, Eglin AFB, FL.
13. Nguyen, D.H., and B. Widrow. 1990. Neural networks for self-learning control systems. *IEEE Control Systems Magazine* 10 (3): 18–23.
14. Barto, A. 1984. Neuron-like adaptive elements that can solve difficult learning control-problems. *Behavioural Processes* 9 (1).
15. Sanner, R.M., and J.-J.E. Slotine. 1991. Gaussian networks for direct adaptive control. In *IEEE American Control Conference*, 2153–2159.

6 Finite-Time LQR and SDRE for Satellite Formation Flying

Even though the results of the infinite-time LQR, SDRE, DI approaches are seemingly easier to understand and implement and can lead to acceptable results with appropriate tuning, it is very important to understand that such approaches are *not strongly recommended* for satellite formation flying problems in general. This is because in infinite-time formulations, the error is usually driven to zero *asymptotically*. However, as one can see from Chapter 2, the satellite formation flying problem is fundamentally a problem where one should ensure formation flying in two neighboring 'orbits' (unlike the formation flying of aerial vehicles). Hence, the right problem formulation should ensure that the relative desired position and velocity vectors are achieved at a '*particular time*' (not earlier, not later). Once that is achieved, from that time onward, the deputy satellite remains in the desired orbit with respect to the chief satellite. Hence, such a problem formulation should ideally be done under the '*finite-time*' optimal control paradigm instead. To address such finite-time terminal constraints problems, fortunately a few advanced techniques are also available in the literature. The present and next chapters deal with a few of these techniques, and demonstrate the suitability and applicability of these methods for satellite formation flying problems. In this chapter, an overview of the finite-time LQR and SDRE techniques is presented, followed by their usage for the satellite formation flying application.

6.1 Finite-Time LQR: Generic Theory

In the finite-time LQR approach, with hard terminal constraint on the state vector, the following cost function is minimized

$$J = \frac{1}{2}\int_{t_0}^{t_f} \left(\mathbf{x}^T Q\mathbf{x} + \mathbf{u}^T R\mathbf{u}\right) dt \tag{6.1}$$

subject to the state equation

$$\dot{\mathbf{x}} = A\mathbf{x} + B\mathbf{u} \tag{6.2}$$

while imposing the following hard constraint on the final states

---

Electronic supplementary material The online version of this chapter (https://doi.org/10.1007/978-981-15-9631-5_6) contains supplementary material, which is available to authorized users.

S. Mathavaraj and R. Padhi, *Satellite Formation Flying*,
https://doi.org/10.1007/978-981-15-9631-5_6

$$\mathbf{x}(t_f) = 0 \tag{6.3}$$

It is also assumed here that the initial condition information $\mathbf{x}(t_0)$ is known.

To proceed with the solution, following the classical optimal control theory [1, 2], first the augmented cost function can be written as

$$\bar{J} = \underbrace{\boldsymbol{\nu}_f^T \mathbf{x}(t_f)}_{\Psi} + \frac{1}{2}\int_{t_0}^{t_f} \left(\mathbf{x}^T Q \mathbf{x} + \mathbf{u}^T R \mathbf{u}\right) dt \tag{6.4}$$

where $\boldsymbol{\nu}_f$ is a Lagrange multiplier (a free variable) and $\Psi \triangleq \boldsymbol{\nu}_f^T \mathbf{x}(t_f)$. Next, using the standard procedure, the Hamiltonian H is constructed as follows

$$H = \frac{1}{2}\left(\mathbf{x}^T Q \mathbf{x} + \mathbf{u}^T R \mathbf{u}\right) + \boldsymbol{\lambda}^T(t)\left(A\mathbf{x} + B\mathbf{u}\right) \tag{6.5}$$

where $\boldsymbol{\lambda}(t)$ is the costate variable. After that, the necessary conditions of optimality [1, 2] lead to the following equations as part of the necessary conditions

$$\dot{\mathbf{x}} = \frac{\partial H}{\partial \boldsymbol{\lambda}} = A\mathbf{x} + B\mathbf{u} \tag{6.6}$$

$$\dot{\boldsymbol{\lambda}} = -\frac{\partial H}{\partial \mathbf{x}} = -Q\mathbf{x} - A^T \boldsymbol{\lambda} \tag{6.7}$$

$$\mathbf{u} = -R^{-1} B^T \boldsymbol{\lambda} \tag{6.8}$$

where Eq. (6.8) comes from the condition $[\partial H / \partial \mathbf{u}] = 0$. Substituting Eq. (6.8) in Eq. (6.7) results in

$$\dot{\mathbf{x}} = A\mathbf{x} - BR^{-1}B^T \boldsymbol{\lambda} \tag{6.9}$$

Now, one can combine the state equation (6.9) and costate equation (6.7) to write

$$\begin{bmatrix} \dot{\mathbf{x}} \\ \dot{\boldsymbol{\lambda}} \end{bmatrix} = \begin{bmatrix} A & -BR^{-1}B^T \\ -Q & -A^T \end{bmatrix} \begin{bmatrix} \mathbf{x} \\ \boldsymbol{\lambda} \end{bmatrix} = A_a \begin{bmatrix} \mathbf{x} \\ \boldsymbol{\lambda} \end{bmatrix} \tag{6.10}$$

where

$$A_a \triangleq \begin{bmatrix} A & -BR^{-1}B^T \\ -Q & -A^T \end{bmatrix} \tag{6.11}$$

Next, following the linear systems theory [1, 2], the solution of Eq. (6.10) can be represented as

$$\begin{bmatrix} \mathbf{x}(t) \\ \boldsymbol{\lambda}(t) \end{bmatrix} = \left[\varphi(t, t_f)\right] \begin{bmatrix} \mathbf{x}(t_f) \\ \boldsymbol{\lambda}(t_f) \end{bmatrix} \tag{6.12}$$

where $\varphi(t, t_f)$ is known as the state transition matrix, which can be written in the partitioned form as follows

$$\left[\varphi(t, t_f)\right]_{2n \times 2n} = \left[\begin{array}{c:c} \varphi_{11}(t, t_f)_{n \times n} & \varphi_{12}(t, t_f)_{n \times n} \\ \hdashline \varphi_{21}(t, t_f)_{n \times n} & \varphi_{22}(t, t_f)_{n \times n} \end{array}\right] \tag{6.13}$$

Evaluating the transversality boundary condition leads to

$$\boldsymbol{\lambda}(t_f) = \left[\frac{\partial \Psi}{\partial \mathbf{x}}\right]_{t_f} = \boldsymbol{\nu}_f \tag{6.14}$$

Next, substituting the state boundary condition Eq. (6.3) and costate boundary condition Eq. (6.14) at t_f in Eq. (6.12) results in

$$\begin{bmatrix} \mathbf{x}(t) \\ \boldsymbol{\lambda}(t) \end{bmatrix} = \begin{bmatrix} \varphi_{11}(t, t_f) & \varphi_{12}(t, t_f) \\ \varphi_{21}(t, t_f) & \varphi_{22}(t, t_f) \end{bmatrix} \begin{bmatrix} 0 \\ \boldsymbol{\nu}_f \end{bmatrix} = \begin{bmatrix} \varphi_{12}(t, t_f) \\ \varphi_{22}(t, t_f) \end{bmatrix} \boldsymbol{\nu}_f \tag{6.15}$$

Now, substituting Eq. (6.15) in Eq. (6.10) and canceling out $\boldsymbol{\nu}_f$ from both sides results in the following system dynamics for the relevant portion of the state transition matrix

$$\begin{bmatrix} \dot{\varphi}_{12}(t, t_f) \\ \dot{\varphi}_{22}(t, t_f) \end{bmatrix} = A_a \begin{bmatrix} \varphi_{12}(t, t_f) \\ \varphi_{22}(t, t_f) \end{bmatrix} \tag{6.16}$$

Moreover, using the state boundary condition Eq. (6.3) and costate boundary condition Eq. (6.14), the following terminal condition of the STM at t_f is obtained.

$$\mathbf{x}(t_f) = \varphi_{12}(t_f, t_f)\boldsymbol{\nu}_f = 0 \tag{6.17}$$

$$\boldsymbol{\lambda}(t_f) = \varphi_{22}(t_f, t_f)\boldsymbol{\nu}_f = \boldsymbol{\nu}_f \tag{6.18}$$

However, since $\boldsymbol{\nu}_f$ is free, it cannot be [0]. Hence, from Eq. (6.17), we can write

$$\varphi_{12}(t_f, t_f) = [0]_{n \times n} \tag{6.19}$$

Similarly, from Eq. (6.18), we can write

$$\varphi_{22}(t_f, t_f) = [I]_{n \times n} \tag{6.20}$$

In general, one has to numerically evaluate the Eq. (6.16) using the boundary condition in Eqs. (6.19) and (6.20) to compute the state transition matrix solution. After the computation of STM solution, from upper half of Eq. (6.15) at t_0, we can compute

$$\boldsymbol{\nu}_f = \varphi_{12}^{-1}(t_0, t_f)\mathbf{x}(t_0) \tag{6.21}$$

and then from the lower half of Eq. (6.15) the costate is computed as

$$\boldsymbol{\lambda}(t) = \varphi_{22}(t, t_f)\boldsymbol{\nu}_f \tag{6.22}$$

Finally, for continuous data $t_0 \to t$, the optimal control is evaluated using Eqs. (6.8), (6.21), (6.22), which can be written as

$$\begin{aligned} \mathbf{u}(t) &= -K(t)\mathbf{x}(t) \\ &= -R^{-1}(\mathbf{x})B^T(\mathbf{x})\varphi_{22}(t, t_f)\varphi_{12}^{-1}(t, t_f)\mathbf{x}(t) \end{aligned} \tag{6.23}$$

It can be mentioned here that the above discussion is generic and holds good for a wide variety of cases. It is also valid for time-varying systems, with only some of the states being constrained at t_f and so on.

However, such a generic approach is avoidable whenever A, B, Q, R matrices are time-invariant, leading to A_a being time-invariant, and the full state vector is constrained at final time $\mathbf{x}(t_f)$. In such a situation, a much simpler approach can be followed, which is discussed next. Whenever A_a is a time-invariant matrix, from the standard results of linear systems theory [1, 2], the state transition matrix can be written as

$$\varphi(t, t_0) \triangleq \left[\begin{array}{c:c} \varphi_{11}(t, t_0) & \varphi_{12}(t, t_0) \\ \hdashline \varphi_{21}(t, t_0) & \varphi_{22}(t, t_0) \end{array}\right] = e^{A_a(t-t_0)} \tag{6.24}$$

Note that the partition matrices $\varphi_{ij}(t, t_0)$ $i, j = 1, 2$ can be extracted from the above definition. With this, one can then write

$$\begin{bmatrix} \mathbf{x}(t) \\ \boldsymbol{\lambda}(t) \end{bmatrix} = \left[\begin{array}{c:c} \varphi_{11}(t, t_0) & \varphi_{12}(t, t_0) \\ \hdashline \varphi_{21}(t, t_0) & \varphi_{22}(t, t_0) \end{array}\right] \begin{bmatrix} \mathbf{x}(t_0) \\ \boldsymbol{\lambda}(t_0) \end{bmatrix} \tag{6.25}$$

From Eq. (6.25), it is obvious that at $t = t_f$ the following relationship holds good

$$\begin{bmatrix} \mathbf{x}(t_f) \\ \boldsymbol{\lambda}(t_f) \end{bmatrix} = \left[\begin{array}{c:c} \varphi_{11}(t_f, t_0) & \varphi_{12}(t_f, t_0) \\ \hdashline \varphi_{21}(t_f, t_0) & \varphi_{22}(t_f, t_0) \end{array}\right] \begin{bmatrix} \mathbf{x}(t_0) \\ \boldsymbol{\lambda}(t_0) \end{bmatrix} \tag{6.26}$$

Hence, from Eq. (6.26), it can be written as

$$\mathbf{x}(t_f) = \varphi_{11}(t_f, t_0)\mathbf{x}(t_0) + \varphi_{12}(t_f, t_0)\boldsymbol{\lambda}(t_0) \tag{6.27}$$

From Eq. (6.27), $\boldsymbol{\lambda}(t_0)$ can be calculated as

$$\boldsymbol{\lambda}(t_0) = \varphi_{12}^{-1}(t_f, t_0)\left[\mathbf{x}(t_f) - \varphi_{11}(t_f, t_0)\mathbf{x}(t_0)\right] \tag{6.28}$$

In other words, the hard constraint information $\mathbf{x}\left(t_f\right)$ and the initial condition information $\mathbf{x}\left(t_0\right)$ are utilized in Eq. (6.28) to compute $\boldsymbol{\lambda}(t_0)$. For continuous data $t_0 \rightarrow t$, optimal control is calculated using Eq. (6.8), which can be expressed as

$$\begin{aligned} \mathbf{u}(t) &= -R^{-1}B^T\boldsymbol{\lambda}(t) \\ &= -R^{-1}B^T\varphi_{12}^{-1}(t_f, t)\left[\mathbf{x}(t_f) - \varphi_{11}(t_f, t)\mathbf{x}(t)\right] \end{aligned} \tag{6.29}$$

6.2 Finite-Time SDRE: Generic Theory

In Section 6.1, the finite-time LQR theory-based guidance is based on the linear approximation of the plant model. The approach obviously fails if this approximation is violated substantially. To address this issue to a limited extent, nonlinear finite-time state-dependent Riccati Equation (F-SDRE) guidance scheme is presented next. The key idea here is to describe the system dynamics in linear-looking state-dependent coefficient (SDC) form and then apply the results from the LQR theory repeatedly online.

It is worth mentioning here that there are several ways of solving a finite-time SDRE problem, three of which are summarized nicely in [3, 4] (an interested reader is strongly advised to refer it). One such method, employing the state transition approach, has been used here and hence is reproduced below for completeness.

In the finite-time SDRE problem, the following cost function is minimized

$$J = \frac{1}{2}\int_{t_0}^{t_f} \left(\mathbf{x}^T Q(\mathbf{x})\mathbf{x} + \mathbf{u}^T R(\mathbf{x})\mathbf{u}\right) dt \tag{6.30}$$

subject to the state equation

$$\dot{\mathbf{x}} = A(\mathbf{x})\mathbf{x} + B(\mathbf{x})\mathbf{u} \tag{6.31}$$

while imposing a hard constraint on the final states

$$\mathbf{x}(t_f) = 0 \tag{6.32}$$

Similar to the discussion in Section 6.1, following the steps outlined in Eqs. (6.4)–(6.9), one can arrive at

$$\begin{bmatrix} \dot{\mathbf{x}} \\ \dot{\boldsymbol{\lambda}} \end{bmatrix} = \begin{bmatrix} A(\mathbf{x}) & -B(\mathbf{x})R^{-1}(\mathbf{x})B^T(\mathbf{x}) \\ -Q(\mathbf{x}) & -A(\mathbf{x})^T \end{bmatrix} \begin{bmatrix} \mathbf{x} \\ \boldsymbol{\lambda} \end{bmatrix} = A_a \begin{bmatrix} \mathbf{x} \\ \boldsymbol{\lambda} \end{bmatrix} \tag{6.33}$$

where

$$A_a = \begin{bmatrix} A(\mathbf{x}) & -B(\mathbf{x})R^{-1}(\mathbf{x})B^T(\mathbf{x}) \\ -Q(\mathbf{x}) & -A(\mathbf{x})^T \end{bmatrix}$$

Note that the system matrix A_a is state-dependent and hence time-varying. However, following the linear systems theory for time-varying systems [1, 2], the solution of Eq. (6.33) can still be represented as

$$\begin{bmatrix} \mathbf{x}(t) \\ \boldsymbol{\lambda}(t) \end{bmatrix} = [\varphi(t, t_f)] \begin{bmatrix} \mathbf{x}(t_f) \\ \boldsymbol{\lambda}(t_f) \end{bmatrix} \tag{6.34}$$

where $\varphi(t, t_f)$ is the state transition matrix. Unfortunately, since A_a is time-varying, a closed-form expression is not possible. However, following the concept discussed in Section 6.1, the system dynamics for the relevant portion of the state transition matrix can be derived as

$$\begin{bmatrix} \dot{\varphi}_{12}(t, t_f) \\ \hdashline \dot{\varphi}_{22}(t, t_f) \end{bmatrix} = A_a \begin{bmatrix} \varphi_{12}(t, t_f) \\ \hdashline \varphi_{22}(t, t_f) \end{bmatrix} \tag{6.35}$$

with the following boundary condition

$$\varphi_{12}(t_f, t_f) = [0]_{n\times n} \tag{6.36}$$

$$\varphi_{22}(t_f, t_f) = [I]_{n\times n} \tag{6.37}$$

Hence, to make use of this approach, one has to numerically evaluate the Eq. (6.35) using the boundary condition in Eqs. (6.36) and (6.37) to compute the state transition matrix solution. After the computation of STM solution, the costate is computed using

$$\boldsymbol{\lambda}(t) = \varphi_{22}(t, t_f)\varphi_{12}^{-1}(t_0, t_f)\mathbf{x}(t_0) \tag{6.38}$$

and finally for continuous data $t_o \to t$, the optimal control is evaluated using Eqs. (6.8) and (6.38), which can be expressed as

$$\begin{aligned}\mathbf{u}(t) &= -K(t)\mathbf{x}(t)\\ &= -R^{-1}(\mathbf{x})B^T(\mathbf{x})\varphi_{22}(t, t_f)\varphi_{12}^{-1}(t, t_f)\mathbf{x}(t)\end{aligned} \tag{6.39}$$

For more details of this solution approach, one can refer to [1].

Even though the above theory is sound and complete, a numerical solution approach is seldom preferred in the guidance applications. Rather, a closed-form solution is always the preferred choice. A natural curiosity then is to explore the possibility of using the same time-invariant solution of Eq. (6.24) as a close approximation to the actual solution, i.e.,

$$\varphi(t, t_0) \triangleq \begin{bmatrix}\varphi_{11}(t, t_0) & \varphi_{12}(t, t_0)\\ \varphi_{21}(t, t_0) & \varphi_{22}(t, t_0)\end{bmatrix} = e^{A_a(t-t_0)} \tag{6.40}$$

where $\varphi(t, t_0)$ is repeatedly evaluated based on the updated A_a matrix at each time step under the '*quasi-steady approximation*'. We reiterate again that Eq. (6.40) is valid as long as A_a is time-invariant. The matrices A, B, Q, R are time-varying in general, because of which the matrix A_a also becomes time-varying. Due to this fact, the closed-form solution for $\varphi(t, t_0)$ in Eq. (6.40) is not valid in general.

To justify this '*quasi-steady approximation*' approach for the problem under discussion, the eigenvalues of the A_a matrix (which play a key role in the state solution) are computed and plotted in Figs. 6.1 and 6.2. It is pretty much obvious from these plots that all the eigenvalues are indeed approximately constant (for all practical purposes, one can claim that those are nothing but constants). This justifies the computation of the STM using Eq. (6.40).

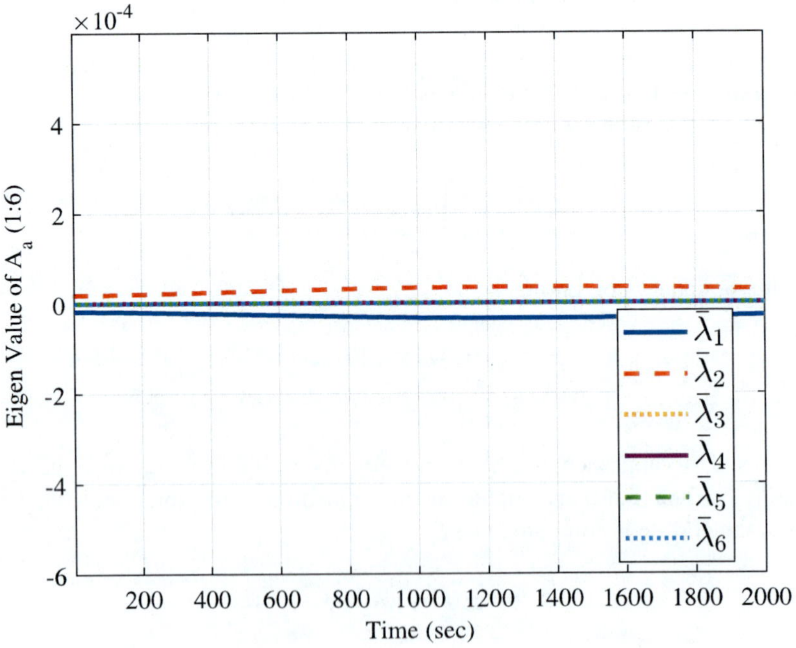

Fig. 6.1 Eigenvalue of A_a (1:6) corresponding to satellite formation flying

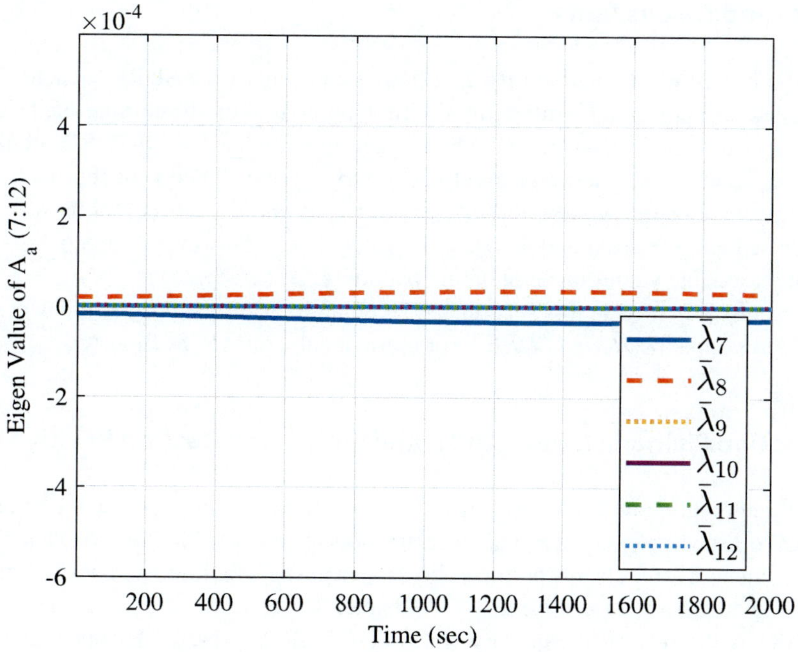

Fig. 6.2 Eigenvalue of A_a (7:12) corresponding to satellite formation flying

Hence, following the similar steps as in Section 6.1, using Eq. (6.40), one can then write

$$\begin{bmatrix} \mathbf{x}(t) \\ \boldsymbol{\lambda}(t) \end{bmatrix} = \left[\begin{array}{c:c} \varphi_{11}(t, t_0) & \varphi_{12}(t, t_0) \\ \hdashline \varphi_{21}(t, t_0) & \varphi_{22}(t, t_0) \end{array}\right] \begin{bmatrix} \mathbf{x}(t_0) \\ \boldsymbol{\lambda}(t_0) \end{bmatrix} \tag{6.41}$$

From Eq. (6.41), it is obvious that at $t = t_f$ the following relationship holds good

$$\begin{bmatrix} \mathbf{x}(t_f) \\ \boldsymbol{\lambda}(t_f) \end{bmatrix} = \left[\begin{array}{c:c} \varphi_{11}(t_f, t_0) & \varphi_{12}(t_f, t_0) \\ \hdashline \varphi_{21}(t_f, t_0) & \varphi_{22}(t_f, t_0) \end{array}\right] \begin{bmatrix} \mathbf{x}(t_0) \\ \boldsymbol{\lambda}(t_0) \end{bmatrix} \tag{6.42}$$

Hence, from Eq. (6.42), it can be written as

$$\mathbf{x}(t_f) = \varphi_{11}(t_f, t_0)\mathbf{x}(t_0) + \varphi_{12}(t_f, t_0)\boldsymbol{\lambda}(t_0) \tag{6.43}$$

From Eq. (6.43), $\boldsymbol{\lambda}(t_0)$ can be calculated as

$$\boldsymbol{\lambda}(t_0) = \varphi_{12}^{-1}(t_f, t_0)\left[\mathbf{x}(t_f) - \varphi_{11}(t_f, t_0)\mathbf{x}(t_0)\right] \tag{6.44}$$

In other words, the hard constraint information $\mathbf{x}\left(t_f\right)$ and the initial condition information $\mathbf{x}\left(t_0\right)$ are utilized in Eq. (6.44) to compute $\boldsymbol{\lambda}(t_0)$. For continuous data $t_o \to t$, the final expression of this suboptimal controller (under the quasi-steady approximation) using Eq. (6.8) can be written as

$$\begin{aligned} \mathbf{u}(t) &= -R^{-1}(\mathbf{x})B^T(\mathbf{x})\boldsymbol{\lambda}(t) \\ &= -R^{-1}(\mathbf{x})B^T(\mathbf{x})\varphi_{12}^{-1}(t_f, t)\left[\mathbf{x}(t_f) - \varphi_{11}(t_f, t)\mathbf{x}(t)\right] \end{aligned} \tag{6.45}$$

6.3 Results and Discussions

Results from the finite-time LQR (denoted as F-LQR) and finite-time SDRE (denoted as F-SDRE) are now generated and presented. Unlike infinite-time approach, the finite-time method attempts to achieve the final states as 'hard constraints'. Once again for completeness, F-SDRE simulation results for both SDC_1 and SDC_2 formulations (refer to Section 3.4) are included with appropriate comments. For the results presented here, without loss of generality, weighting matrices on the state and control vectors are selected as $Q = [0]_6$ and $R = 10^9 I_3$, respectively. However, a practicing engineer can select different reasonable values of these tuning parameters if he/she wishes to.

Simulation results provided in this section can be obtained from the program files provided in the folder named '*Finite-time: LQR and SDRE*'. For more details, refer to Section A.4.

6.3.1 Formation Flying in Elliptic Orbit and with Small Desired Relative Distance

Without loss of generality, out of all cases discussed in Chapter 2, the case study pertaining to chief orbit in elliptical orbit with deputy at a smaller relative distance has been selected. This is to show that the formulation due to SDC_2 always performs better. The initial condition and desired final condition for this case study are selected from Table 2.3 of Chapter 2.

Figure 6.3 shows the resulting trajectory states of F-LQR and F-SDRE approaches considering eccentricity $e = 0.05$ (refer to Table 2.3) for chief satellite. From this figure, it can be seen that SDC_2-based F-SDRE performs much better in achieving the intended desired orbit, whereas F-LQR as well as SDC_1-based F-SDRE controllers deviate away and hence do not perform satisfactorily. This is obvious because the SDC_2 formulation accounts for the nonlinearity due to eccentric better (see Section 3.4 for details). Since the control histories appear similar, the position and velocity error histories of the guidance techniques (shown in Figs. 6.4 and 6.5, respectively) look similar and the errors seem to approach zero at final time. Figure 6.6 gives the associated control histories for the guidance techniques discussed.

However, the accuracy with which the terminal condition is achieved is of much significance in satellite formation flying problem. The reason is the orbit correcting formation flying controller (i.e., the guidance action) is not supposed to act beyond t_f and both satellites are supposed to fly in their respective orbits autonomously. Hence, if high level of accuracy is not achieved at t_f, as the time elapses beyond t_f, the deputy satellite will deviate away from the desired configuration with respect to the chief satellite. Keeping this in mind, the resulting terminal accuracy was collected at t_f and tabulated in Table 6.1. This table shows that the state errors at final time of F-LQR and SDC_1-based F-SDRE formulation are much larger as compared to the SDC_2-based F-SDRE formulation.

Table 6.1 Finite-time formulation terminal error ($\rho_i = 1$ km, $\rho_f = 5$ km, $e = 0.05$)

State error	F-LQR	SDC_1	SDC_2
Δx (km)	-2.35×10^{-1}	-2.36×10^{-1}	1.75×10^{-3}
$\Delta \dot{x}$ (km/s)	-2.75×10^{-4}	-2.76×10^{-4}	5.29×10^{-6}
Δy (km)	2.02×10^{-2}	1.94×10^{-2}	3.78×10^{-3}
$\Delta \dot{y}$ (km/s)	1.12×10^{-4}	1.11×10^{-4}	-4.95×10^{-5}
Δz (km)	7.54×10^{-3}	7.34×10^{-3}	7.34×10^{-3}
$\Delta \dot{z}$ (km/s)	2.01×10^{-5}	1.90×10^{-5}	1.88×10^{-5}
Norm	2.36×10^{-1}	2.37×10^{-1}	8.44×10^{-3}

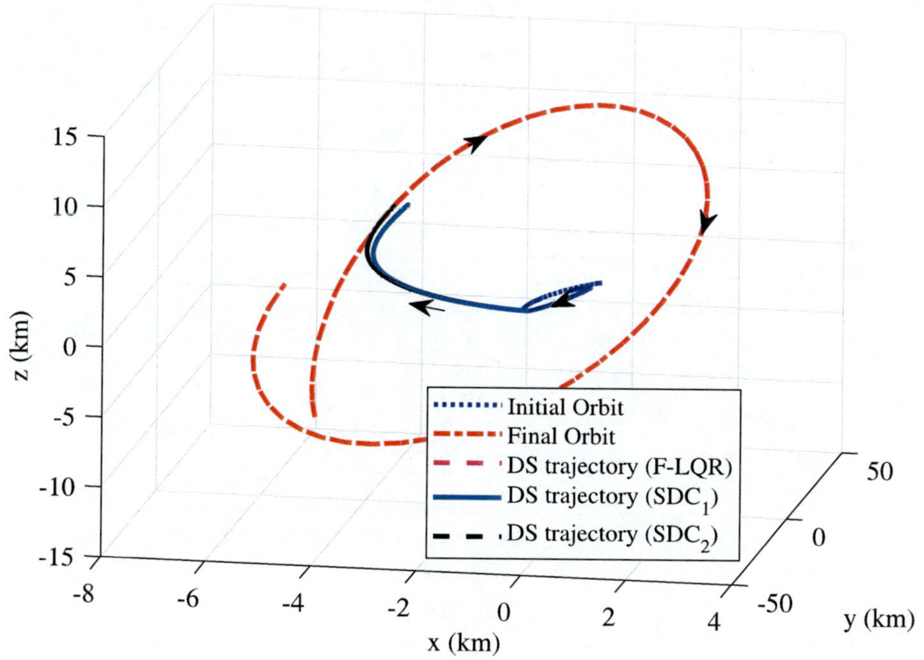

Fig. 6.3 Deputy satellite (DS) trajectory ($\rho_i = 1$ km, $\rho_f = 5$ km, $e = 0.05$)

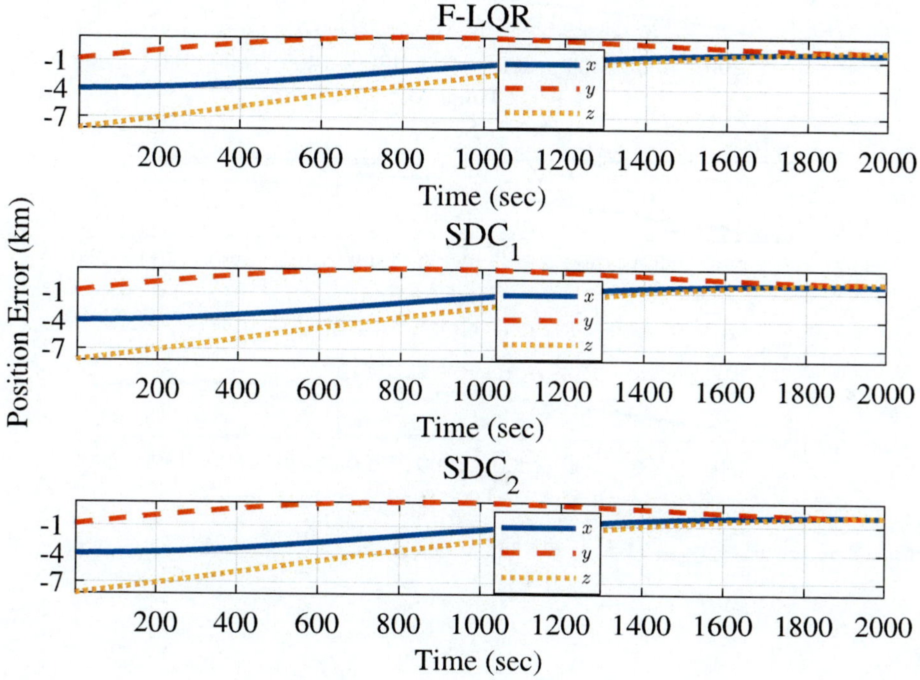

Fig. 6.4 Position error history ($\rho_i = 1$ km, $\rho_f = 5$ km, $e = 0.05$)

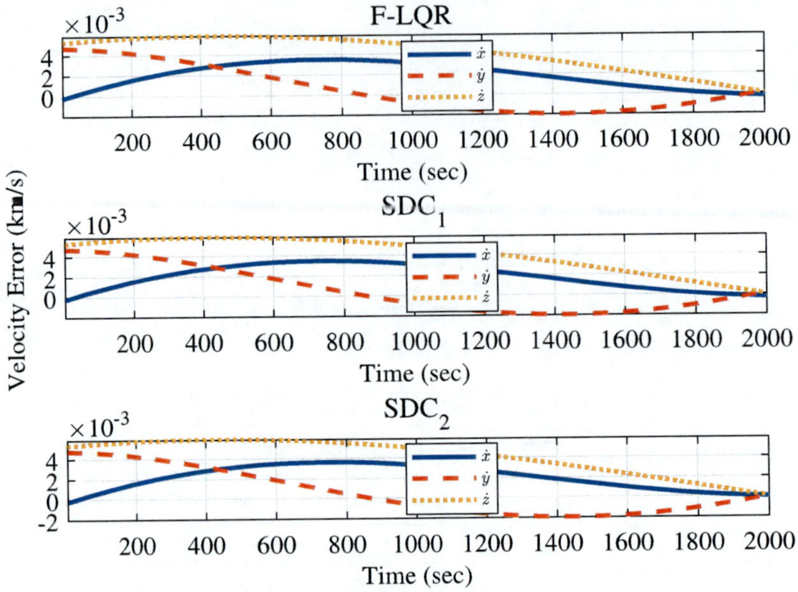

Fig. 6.5 Velocity error history ($\rho_i = 1\,\text{km}$, $\rho_f = 5\,\text{km}$, $e = 0.05$)

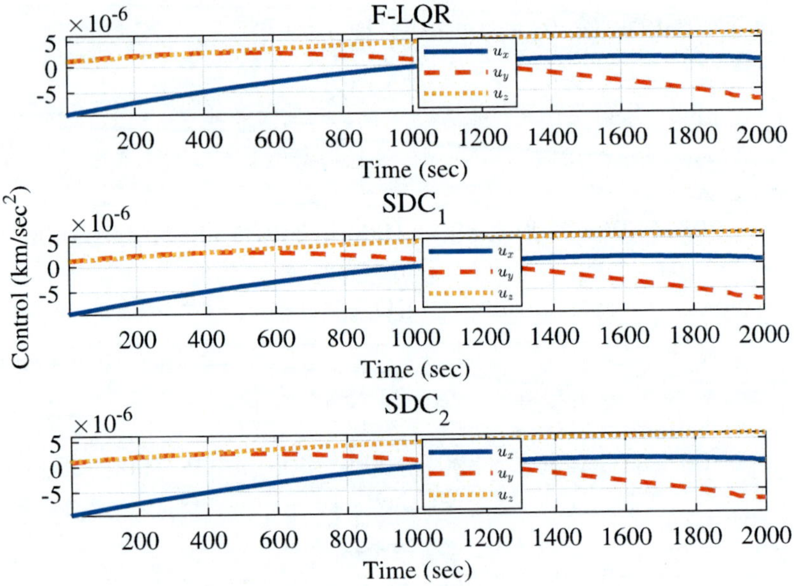

Fig. 6.6 Control history for DS ($\rho_i = 1\,\text{km}$, $\rho_f = 5\,\text{km}$, $e = 0.05$)

Note that F-LQR and SDC_1-based F-SDRE formulations are derived from the relative dynamics under the assumption of circular orbit of the chief satellite. Hence, the nonlinear behavior of the problem due to eccentricity of the chief satellite's orbit is not captured well in these formulations and hence results in higher errors. However, the SDC_2-based F-SDRE formulation retains the nonlinearity of the problem to the maximum extent possible accounting for the eccentricity of the chief satellite's orbit and hence results in better performance. An interested reader is referred to Section 3.4 for details.

6.4 Summary

In this chapter, we presented an overview of the finite-time LQR and finite-time SDRE control design techniques using the state transition approach. Next, with numerical results, it is shown that F-LQR and F-SDRE approach using SDC_1 do not perform satisfactorily if the chief satellite is in eccentric orbit, where as the SDC_2-based F-SDRE approach performs much better and hence it is strongly recommended.

References

1. Bryson, A.E. and Y.C. Ho. 1975. *Applied optimal control*. Hemisphere Publishing Corporation.
2. Naidu, D.S. 2002. *Optimal control systems*. CRC Press.
3. Korayem, M., and S. Nekoo. 2015. Finite-time state-dependent Riccati equation for time-varying nonaffine systems: Rigid and flexible joint manipulator control. *ISA Transactions* 54: 125–144.
4. Khamis, A., and D. Naidu. 2013. Nonlinear optimal tracking using finite-horizon state dependent Riccati equation (SDRE). In *Proceedings of the 4th International Conference on Circuits, Systems, Control, Signals*, 37–42.

Check for updates

Model Predictive Static Programming

7

In Chapters. 3 and 6, a nonlinear suboptimal control design technique known as the SDRE technique has been successfully employed (both in infinite-time and finite-time frameworks) for the satellite formation flying problem. However, the method requires that the plant model to be expressed in the linear-looking state-dependent coefficient form, which is a nontrivial task in general. Moreover, because of the usage of linear optimal control theory, the resulting solution turns out to be suboptimal. In order to overcome such issues, an alternate optimal control design approach, known as the model predictive static programming (MPSP), is proposed in this chapter. This fairly recent approach solves the original nonlinear optimal control problem in a computationally efficient manner without introducing any transformation or approximation. Because of its fast convergence with less computational demand, it is suitable for implementation in the onboard processors as well. The generic technique is discussed first, followed by its usage for controlling the problem of satellite formation flying problem. Both discrete and continuous time versions of the MPSP technique are presented here for completeness.

7.1 Discrete MPSP: Generic Theory

Consider the following nonlinear system

$$\dot{\mathbf{x}} = f\left(\mathbf{x}, \mathbf{u}\right) \tag{7.1}$$

$$\mathbf{y} = h\left(\mathbf{x}\right) \tag{7.2}$$

where $\mathbf{x} \in \Re^n$, $\mathbf{u} \in \Re^m$, $\mathbf{y} \in \Re^p$ are the state, control, and output vectors, respectively. The primary objective is to obtain a suitable control history $\mathbf{u}$ so that the output $\mathbf{y}(t_f)$ at the fixed final time t_f goes to a desired value $\mathbf{y}^*(t_f)$, i.e., $\mathbf{y}(t_f) \rightarrow \mathbf{y}^*(t_f)$ with minimum control effort. It is assumed that the initial condition for the nonlinear system Eq. (7.1), i.e., $\mathbf{x}(t_0) = \mathbf{x}_0$ is known.

Next, utilizing appropriate discretization formula [1], the discretized state dynamics and output equations can be written as

$$\mathbf{x}_{k+1}^i = F_k\left(\mathbf{x}_k^i, \mathbf{u}_k^i\right) \tag{7.3}$$

Electronic supplementary material The online version of this chapter (https://doi.org/10.1007/978-981-15-9631-5_7) contains supplementary material, which is available to authorized users.

S. Mathavaraj and R. Padhi, *Satellite Formation Flying*,
https://doi.org/10.1007/978-981-15-9631-5_7

$$\mathbf{y}_{\tilde{k}}^i = H_{\tilde{k}}\left(\mathbf{x}_{\tilde{k}}^i\right) \tag{7.4}$$

where $k = 1, \ldots, N-1$, and $\tilde{k} = 1, \ldots, N$ are the discretized time steps in the ith iteration, where $i = 1, 2, \ldots$ represents the iteration index. The primary objective is to come up with a suitable control history $\mathbf{u}_k^i$, $k = 1, 2, \ldots, N-1$, so that the output at the final time step $\mathbf{y}_N$ reaches a desired value $\mathbf{y}_N^*$ (i.e., $\mathbf{y}_N^i \rightarrow \mathbf{y}_N^*$). In addition, the aim is to achieve this task with minimum control effort. It has to be noted that for continuous-time systems, the system dynamics should be first written in a discretized manner as in Eqs. (7.3) and (7.4) (for example, using Euler integration approximation) to implement the algorithm.

Like a typical optimal control solution approach, the idea is to predict the system behavior with the most recent update of the control history (starting from an initial guess solution) and then to quickly update it with the error information available at the final time and the process is repeated so as to drive $\mathbf{y}_N^i \rightarrow \mathbf{y}_N^*$ with as less number of iterations as possible. Note that like other algorithms, a fairly good guess history is also recommended to begin the iteration process so that the algorithm converges quickly. However, the method to obtain a good guess control history is obviously problem specific. With the application of a guess history, however, the objective is not expected to be met. Hence, there is a need to improve this solution, which is done in an iterative manner based on the following mathematical development.

To proceed with the mathematical development, it is assumed that there exists some $\mathbf{x}_N^*$ for which the same function in Eq. (7.4), when evaluated at $\mathbf{x}_N^*$, will result in the $\mathbf{y}_N^*$; i.e., one can write

$$\mathbf{y}_N^* = H_N\left(\mathbf{x}_N^*\right) \tag{7.5}$$

Note that $\mathbf{x}_N^*$ can be interpreted as the 'desired value' for $\mathbf{x}_N^i$ such that when $\mathbf{x}_N^i$ is driven to $\mathbf{x}_N^*$, the objective of $\mathbf{y}_N^i$ being driven to $\mathbf{y}_N^*$ is met. Next, expanding $\mathbf{y}_N^*$ in Taylor series about $\mathbf{x}_N^i$ and neglecting higher order terms, one gets

$$\mathbf{y}_N^* \approx \mathbf{y}_N^i + \left[\frac{\partial \mathbf{y}_N^i}{\partial \mathbf{x}_N^i}\right]_{\mathbf{x}_N^i} \left(\mathbf{x}_N^* - \mathbf{x}_N^i\right) \tag{7.6}$$

which gives

$$\left(\mathbf{y}_N^* - \mathbf{y}_N^i\right) \approx \left[\frac{\partial \mathbf{y}_N^i}{\partial \mathbf{x}_N^i}\right]_{\mathbf{x}_N^i} \left(\mathbf{x}_N^* - \mathbf{x}_N^i\right) \tag{7.7}$$

Next, define the error in the output and the state at step $k = N$ as $\Delta\mathbf{y}_N^i \triangleq \left(\mathbf{y}_N^* - \mathbf{y}_N^i\right)$ and $\Delta\mathbf{x}_N^i \triangleq \left(\mathbf{x}_N^* - \mathbf{x}_N^i\right)$, respectively. By assuming the errors $\Delta\mathbf{y}_N^i$ and $\Delta\mathbf{x}_N^i$ to be small, Eq. (7.7) can be rewritten as

$$d\mathbf{y}_N^i = \left[\frac{\partial \mathbf{y}_N^i}{\partial \mathbf{x}_N^i}\right] d\mathbf{x}_N^i \tag{7.8}$$

where small $\Delta\mathbf{y}_N^i$ and $\Delta\mathbf{x}_N^i$ are denoted as $d\mathbf{y}_N^i$ and $d\mathbf{x}_N^i$, respectively. However, from Eq. (7.3), one can write

$$\mathbf{x}_N^i = F_{N-1}\left(\mathbf{x}_{N-1}^i, \mathbf{u}_{N-1}^i\right) \tag{7.9}$$

Similarly $\mathbf{x}_N^*$ can be written as

$$\mathbf{x}_N^* = F_{N-1}\left(\mathbf{x}_{N-1}^*, \mathbf{u}_{N-1}^*\right) \tag{7.10}$$

where $\mathbf{x}_{N-1}^*$ and $\mathbf{u}_{N-1}^*$ are the 'desired values' of the state and control at $k = N - 1$ respectively. Next, subtracting Eq. (7.10) from Eq. (7.9), expanding $F_{N-1}\left(\mathbf{x}_{N-1}^*, \mathbf{u}_{N-1}^*\right)$ in Taylor series about $\left(\mathbf{x}_{N-1}^i, \mathbf{u}_{N-1}^i\right)$ and keeping only the first-order terms leads to

$$\begin{aligned}\left(\mathbf{x}_N^* - \mathbf{x}_N^i\right) &= F_{N-1}\left(\mathbf{x}_{N-1}^*, \mathbf{u}_{N-1}^*\right) - F_{N-1}\left(\mathbf{x}_{N-1}^i, \mathbf{u}_{N-1}^i\right) \\ &\approx \left[\frac{\partial F_{N-1}}{\partial \mathbf{x}_{N-1}^i}\right]\left(\mathbf{x}_{N-1}^* - \mathbf{x}_{N-1}^i\right) + \left[\frac{\partial F_{N-1}}{\partial \mathbf{u}_{N-1}^i}\right]\left(\mathbf{u}_{N-1}^* - \mathbf{u}_{N-1}^i\right)\end{aligned} \tag{7.11}$$

By defining $\Delta\mathbf{x}_{N-1}^i \triangleq \left(\mathbf{x}_{N-1}^* - \mathbf{x}_{N-1}^i\right)$ and $\Delta\mathbf{u}_{N-1}^i \triangleq \left(\mathbf{u}_{N-1}^* - \mathbf{u}_{N-1}^i\right)$, Eq. (7.11) can be rewritten as

$$\Delta\mathbf{x}_N^i \approx \left[\frac{\partial F_{N-1}}{\partial \mathbf{x}_{N-1}^i}\right]\Delta\mathbf{x}_{N-1}^i + \left[\frac{\partial F_{N-1}}{\partial \mathbf{u}_{N-1}^i}\right]\Delta\mathbf{u}_{N-1}^i \tag{7.12}$$

With the assumption that $\Delta\mathbf{x}_N^i$, $\Delta\mathbf{x}_{N-1}^i$, and $\Delta\mathbf{u}_{N-1}^i$ are small, they can be denoted as $d\mathbf{x}_N^i$, $d\mathbf{x}_{N-1}^i$, and $d\mathbf{u}_{N-1}^i$, respectively, and hence Eq. (7.12) can be written as

$$d\mathbf{x}_N^i = \left[\frac{\partial F_{N-1}}{\partial \mathbf{x}_{N-1}^i}\right]d\mathbf{x}_{N-1}^i + \left[\frac{\partial F_{N-1}}{\partial \mathbf{u}_{N-1}^i}\right]d\mathbf{u}_{N-1}^i \tag{7.13}$$

In general, for any time step k, defining $\Delta\mathbf{x}_k^i \triangleq \left(\mathbf{x}_k^* - \mathbf{x}_k^i\right)$, $\Delta\mathbf{u}_k^i \triangleq \left(\mathbf{u}_k^* - \mathbf{u}_k^i\right)$ and following exactly the same argument, one can write

$$\Delta\mathbf{x}_{k+1}^i \approx \left[\frac{\partial F_k}{\partial \mathbf{x}_k^i}\right]\Delta\mathbf{x}_k^i + \left[\frac{\partial F_k}{\partial \mathbf{u}_k^i}\right]\Delta\mathbf{u}_k^i \tag{7.14}$$

where $\Delta\mathbf{x}_k^i$ and $\Delta\mathbf{u}_k^i$ are the error of state and control at time step k, respectively. With the assumption that $\Delta\mathbf{x}_{k+1}^i$, $\Delta\mathbf{x}_k^i$ and $\Delta\mathbf{u}_k^i$ are small, they can be denoted as $d\mathbf{x}_{k+1}^i$, $d\mathbf{x}_k^i$ and $d\mathbf{u}_k^i$, respectively, and Eq. (7.14) can be written as

$$d\mathbf{x}_{k+1}^i = \left[\frac{\partial F_k}{\partial \mathbf{x}_k^i}\right]d\mathbf{x}_k^i + \left[\frac{\partial F_k}{\partial \mathbf{u}_k^i}\right]d\mathbf{u}_k^i \tag{7.15}$$

Note that the definitions of $\Delta\mathbf{u}_k^i$ gives the following iteration logic for the control history update

$$\mathbf{u}_k^{i+1} = \mathbf{u}_k^* = \mathbf{u}_k^i + \Delta\mathbf{u}_k^i \tag{7.16}$$

where $\mathbf{u}_k^i$ is the existing value and $\mathbf{u}_k^{i+1}$ is the updated value of the control at time step k. However, since computation of $\Delta\mathbf{u}_k^i$ is difficult, the real control history update is carried out with the small error approximation, i.e., with the approximation $d\mathbf{u}_k^i \approx \Delta\mathbf{u}_k^i$ and the control update is carried out as follows:

$$\mathbf{u}_k^{i+1} \approx \mathbf{u}_k^i + d\mathbf{u}_k^i \tag{7.17}$$

Next, the expression for $d\mathbf{u}_k^i$ needs to be derived in a systematic manner. For this, using Eq. (7.15), equation (7.8) can now be rewritten as

$$d\mathbf{y}_N^i = \left[\frac{\partial \mathbf{y}_N^i}{\partial \mathbf{x}_N^i}\right]\left[\frac{\partial F_{N-1}}{\partial \mathbf{x}_{N-1}^i}\right] d\mathbf{x}_{N-1}^i + \left[\frac{\partial \mathbf{y}_N^i}{\partial \mathbf{x}_N^i}\right]\left[\frac{\partial F_{N-1}}{\partial \mathbf{u}_{N-1}^i}\right] d\mathbf{u}_{N-1}^i \tag{7.18}$$

$$\begin{aligned} &= \left[\frac{\partial \mathbf{y}_N^i}{\partial \mathbf{x}_N^i}\right]\left[\frac{\partial F_{N-1}}{\partial \mathbf{x}_{N-1}^i}\right]\left(\left[\frac{\partial F_{N-2}}{\partial \mathbf{x}_{N-2}^i}\right] d\mathbf{x}_{N-2}^i + \left[\frac{\partial F_{N-2}}{\partial \mathbf{u}_{N-2}^i}\right] d\mathbf{u}_{N-2}^i\right) \\ &\quad + \left[\frac{\partial \mathbf{y}_N^i}{\partial \mathbf{x}_N^i}\right]\left[\frac{\partial F_{N-1}}{\partial \mathbf{u}_{N-1}^i}\right] d\mathbf{u}_{N-1}^i \end{aligned} \tag{7.19}$$

Next, $d\mathbf{x}_{N-2}^i$ can be expanded in terms of $d\mathbf{x}_{N-3}^i$, $d\mathbf{u}_{N-3}^i$ and so on. Continuing the process until $k = 1$ and introducing small error approximations, one can write

$$d\mathbf{y}_N^i = A\ d\mathbf{x}_1^i + B_1 d\mathbf{u}_1^i + B_2 d\mathbf{u}_2^i + \cdots + B_{N-1} d\mathbf{u}_{N-1}^i \tag{7.20}$$

where

$$A \triangleq \left[\frac{\partial \mathbf{y}_N^i}{\partial \mathbf{x}_N^i}\right]\left[\frac{\partial F_{N-1}}{\partial \mathbf{x}_{N-1}^i}\right]\cdots\left[\frac{\partial F_1}{\partial \mathbf{x}_1^i}\right] \tag{7.21}$$

and

$$B_k^i \triangleq \left[\frac{\partial \mathbf{y}_N^i}{\partial \mathbf{x}_N^i}\right]\left[\frac{\partial F_{N-1}}{\partial \mathbf{x}_{N-1}^i}\right]\cdots\left[\frac{\partial F_{k+1}}{\partial \mathbf{x}_{k+1}^i}\right]\left[\frac{\partial F_k}{\partial \mathbf{u}_k^i}\right] \quad \forall\, k = 1, 2, \ldots, N-2 \tag{7.22}$$

$$B_{N-1}^i \triangleq \left[\frac{\partial \mathbf{y}_N^i}{\partial \mathbf{x}_N^i}\right]\left[\frac{\partial F_{N-1}}{\partial \mathbf{u}_{N-1}^i}\right] \tag{7.23}$$

Since the initial condition is specified, there is no error in the first term. Hence $d\mathbf{x}_1^i = 0$, and Eq. (7.20) reduces to

$$d\mathbf{y}_N^i = B_1^i d\mathbf{u}_1^i + B_2^i d\mathbf{u}_2^i + \cdots + B_{N-1}^i d\mathbf{u}_{N-1}^i = \sum_{k=1}^{N-1} B_k^i d\mathbf{u}_k^i \tag{7.24}$$

Note that while deriving Eq. (7.24), we have assumed that the control variable at each time step is independent of the previous values of states and/or control. Justification of this assumption comes from the fact that it is a 'decision variable', and hence, independent decision can be taken at any point of time. Here, it can be pointed out that that if one evaluates the sensitivity matrices B_k, $k = 1, \ldots, (N-1)$ from Eq. (7.22) directly, it will be a computationally intensive task (especially when N is high).

However, fortunately it is possible to compute them recursively, which leads to a substantial saving of computational time. The recursive computation can be done as follows

$$\bar{B}^i_{N-1} = \left[\frac{\partial \mathbf{y}_N}{\partial \mathbf{x}_N}\right] \tag{7.25}$$

$$\bar{B}^i_k = \bar{B}^i_{k+1}\left[\frac{\partial F_{k+1}}{\partial \mathbf{x}_{k+1}}\right] \quad (N-2),\ldots,1 \tag{7.26}$$

$$B_k = \bar{B}^i_k\left[\frac{\partial F_k}{\partial \mathbf{u}_k}\right] \quad (N-1),\ldots,1 \tag{7.27}$$

In Eq. (7.24), we have $(N-1)m$ unknowns and p equations. Usually $p \ll (N-1)m$ and hence, it is an under-constrained system of equations. Hence, there is a scope for meeting additional objectives. We take advantage of this opportunity and aim to minimize the following performance index (which represents a 'control minimization' problem)

$$J = \frac{1}{2}\sum_{k=1}^{N-1}\left[\left(\mathbf{u}_k^{i+1}\right)^T R_k \left(\mathbf{u}_k^{i+1}\right)\right] \tag{7.28}$$

$$= \frac{1}{2}\sum_{k=1}^{N-1}\left[\left(\mathbf{u}_k^{i} + d\mathbf{u}_k^{i}\right)^T R_k \left(\mathbf{u}_k^{i} + d\mathbf{u}_k^{i}\right)\right] \tag{7.29}$$

where $\mathbf{u}_k^i, k = 1,\ldots,(N-1)$ represents the previous control history solution and $d\mathbf{u}_k^i$ is the corresponding error in the control history at time step k. The cost function in Eq. (7.29) needs to be minimized subjected to the constraint in Eq. (7.24). Here, $R_k > 0$ (a positive definite matrix) is the weighting matrix at time step k, which needs to be chosen judiciously by the control designer. The selection of such a performance index is motivated by the fact that we are interested in finding a l_2-norm minimizing control history, since $\mathbf{u}_k^{i+1} = \left(\mathbf{u}_k{}^i + d\mathbf{u}_k^i\right)$ is the updated control value at time step k. Equations (7.24) and (7.29) formulate an appropriate constrained static (parametric) optimization problem. Hence, using static optimization theory [2], the augmented cost function is given by

$$\bar{J} = \frac{1}{2}\sum_{k=1}^{N-1}\left[\left(\mathbf{u}_k^{i} + d\mathbf{u}_k^{i}\right)^T R_k \left(\mathbf{u}_k^{i} + d\mathbf{u}_k^{i}\right)\right] + \boldsymbol{\lambda}^{i\,T}\left(d\mathbf{y}_N^i - \sum_{k=1}^{N-1} B_k^i d\mathbf{u}_k^i\right) \tag{7.30}$$

where $\boldsymbol{\lambda}^i$ is a Lagrange multiplier (adjoint variable). Then, the necessary conditions of optimality for $k = 1,\ldots,N-1$ are given by

$$\frac{\partial \bar{J}}{\partial\left(d\mathbf{u}_k^i\right)} = R_k\left(\mathbf{u}_k^i + d\mathbf{u}_k^i\right) - B_k^{i\,T}\boldsymbol{\lambda}^i = 0 \tag{7.31}$$

$$\frac{\partial \bar{J}}{\partial \boldsymbol{\lambda}^i} = d\mathbf{y}_N^i - \sum_{k=1}^{N-1} B_k^i d\mathbf{u}_k^i = 0 \tag{7.32}$$

Solving for $d\mathbf{u}_k^i$ from Eq. (7.31), we get

$$d\mathbf{u}_k^i = R_k^{-1} B_k^{i\,T}\boldsymbol{\lambda}^i - \mathbf{u}_k^i \tag{7.33}$$

Substituting for $d\mathbf{u}_k^i$ from Eq. (7.33) in Eq. (7.32) leads to

$$A_{\lambda^i}\boldsymbol{\lambda}^i - b_{\lambda^i} = d\mathbf{y}_N^i \tag{7.34}$$

where

$$A_\lambda^i \triangleq \left[\sum_{k=1}^{N-1} B_k^i R_k^{-1} B_k^{i\,T}\right], \quad b_\lambda^i \triangleq \left[\sum_{k=1}^{N-1} B_k^i \mathbf{u}_k^i\right] \tag{7.35}$$

Assuming A_λ^i to be non-singular, the solution for $\boldsymbol{\lambda}^i$ from Eq. (7.34) is given by

$$\boldsymbol{\lambda}^i = A_\lambda^{i\,-1}\left(d\mathbf{y}_N^i + b_\lambda^i\right) \tag{7.36}$$

Using Eq. (7.36) in Eq. (7.33), it leads to

$$d\mathbf{u}_k^i = R_k^{-1} B_k^{i\,T} A_\lambda^{i\,-1}\left(d\mathbf{y}_N^i + b_\lambda^i\right) - \mathbf{u}_k^i \tag{7.37}$$

The updated control at time step $k = 1, \ldots, (N-1)$ is

$$\mathbf{u}_k^{i+1} = \mathbf{u}_k^i + d\mathbf{u}_k^i = R_k^{-1} B_k^{i\,T} A_\lambda^{i\,-1}\left(d\mathbf{y}_N^i + b_\lambda^i\right) \tag{7.38}$$

This control update process needs to be iterated before one arrives at the converged solution, i.e., until $\mathbf{y}_N^i \rightarrow \mathbf{y}_N^*$. In summary (see Fig. 7.1), the algorithm can be implemented with the following steps:

(i) Guess a control history guess $\mathbf{u}_k^i$, $k = 1, 2, \ldots, N-1$.
(ii) Using the known initial condition and the available (most update) control history, propagate the system dynamics in Eq. (7.3) for $k = 1, \ldots, N-1$ and obtain the final value of the state $\mathbf{x}_N^i$. Note that this propagation can be done using a higher order numerical scheme as well (such as using fourth-order Runge–Kutta method).
(iii) Using the states and control at each grid point, compute the sensitivity matrices using the recursive relationship in Eqs. (7.21)–(7.23).
(iv) Using $\mathbf{x}_N^i$, compute $\mathbf{y}_N^i$ and with the availability of $\mathbf{x}_N^*$, compute $d\mathbf{y}_N^i \approx \left(\mathbf{y}_N^* - \mathbf{y}_N^i\right)$. Terminate the algorithm if ΔJ^i and $d\mathbf{y}_N^i$ are very small (i.e., if $J^i - J^{i-1}$ and $\|d\mathbf{y}_N^i\| \,/\, \|\mathbf{y}_N^*\|$ is smaller than a pre-designated small tolerance value) and accept the most recent update of $\mathbf{u}_k^i$, $k = 1, 2, \ldots, N-1$ as the optimal control history. Else continue with the following steps.
(v) Compute A_λ^i and b_λ^i from Eq. (7.35).
(vi) Compute the control update $\mathbf{u}_k^{i+1}$, $k = 1, 2, \ldots, N-1$ from Eq. (7.38).
(vii) Repeat the steps (ii)–(vi). In case of convergence, however, the algorithm stops at step (iv). This converged control vector is denoted as $\mathbf{u}^* = [u_1, \ldots, u_{N-1}]^T$.

From Eq. (7.38), it is clear that the relative magnitude of the control input at various time steps can be handled in a 'soft constraint' manner by adjusting the weight matrices R_k, $k = 1, \ldots, (N-1)$ associated with the cost function. Another important point to note is that the costate (adjoint) variable considered here is not a function of time, unlike the time-varying costate variable considered in the dynamic optimization theory [2]. One may observe that the MPSP technique has been inspired from the philosophies of model predictive control (MPC) [3] and approximate dynamic programming (ADP) [4].

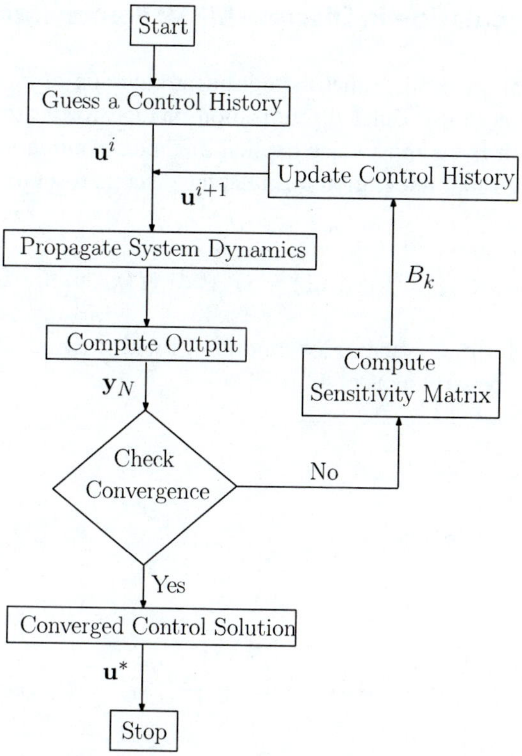

Fig. 7.1 Schematic flowchart for MPSP

Similarity with the MPC technique includes the predictive-corrective nature of the algorithm where the output is predicted and control history is corrected. On the other hand, it also looks similar to ADP in its discrete nature of the dynamic optimization problem formulation and backward propagation of error in the final state (in approximate dynamic programming, the final costate vector is computed from the final state, which is then propagated backwards through the costate equation).

Innovations of the MPSP technique can be attributed to the following facts: (i) in contrast to typical two-point boundary value problems in the optimal control formulations, it rather demands only a static costate vector (that too of the same dimension as the output vector) for the entire control history update, (ii) the costate vector (and hence the control history update) has a symbolic solution, and (iii) the sensitivity matrices that are necessary for obtaining this symbolic solution can be computed recursively. These are the key reasons for its high computational efficiency. Moreover, ideas like 'iteration unfolding' [5], where the control history is updated only a finite number of times in a particular time step, can also be incorporated to enhance the computational efficiency further (at the cost of minor compromise on the optimality of the solution). With the advancement of the computing technology in parallel, it holds good promise for implementation in onboard processors in the near future. Note that MPSP is a technique in itself and, unlike MPC (and other transcription methods), it does not depend on the numerical static (parametric) optimization techniques in the background as long as the problem is free from inequality path constraints. For further details on MPSP, one can refer [6].

7.2 SFF Problem Formulation in Discrete MPSP Framework

The discrete MPSP formulation requires that the nonlinear equation of motion is written in discrete form. One convenient way is to use Euler discretization method for writing the nonlinear equation of motion in discrete form. It is the most basic explicit method for numerical integration of ordinary differential equations. Using Euler integration formula, the discrete form for the system dynamics Eq. (7.1) can be written as

$$\mathbf{x}_{k+1} = F(\mathbf{x}_k, \mathbf{u}_k) = \mathbf{x}_k + \Delta t \;\; f(\mathbf{x}_k, \mathbf{u}_k) \tag{7.39}$$

Using this, the discretized form of equation of motion of relative dynamics of deputy satellite with respect to chief satellite can be written as

$$\mathbf{x}_{k+1} \triangleq \begin{bmatrix} x_{1_{k+1}} \\ x_{2_{k+1}} \\ x_{3_{k+1}} \\ x_{4_{k+1}} \\ x_{5_{k+1}} \\ x_{6_{k+1}} \end{bmatrix} = F(\mathbf{x}_k, \mathbf{u}_k)$$

$$= \begin{bmatrix} x_{1_k} \\ x_{2_k} \\ x_{3_k} \\ x_{4_k} \\ x_{5_k} \\ x_{6_k} \end{bmatrix} + \Delta t \begin{bmatrix} x_{2_k} \\ 2\dot{\nu} x_{4_k} + \ddot{\nu} x_{3_k} + \dot{\nu}^2 x_{1_k} - \frac{\mu}{\gamma_k} x_{1_k} - \frac{\mu}{\gamma_k} r_c + \frac{\mu}{r_c^2} + a_{x_k} + a_{J_{2x}} \\ x_{4_k} \\ -2\dot{\nu} x_{2_k} - \ddot{\nu} x_{1_k} + \dot{\nu}^2 x_{3_k} - \frac{\mu}{\gamma_k} x_{3_k} + a_{y_k} + a_{J_{2y}} \\ x_{6_k} \\ -\frac{\mu}{\gamma_k} x_{5_k} + a_{z_k} + a_{J_{2z}} \end{bmatrix} \tag{7.40}$$

where $\mathbf{x}_k = \left[x_{1_k} \; x_{2_k} \; x_{3_k} \; x_{4_k} \; x_{5_k} \; x_{6_k} \right]^T$, where the subscript $k = 1, \ldots, N$ of all variables stands for the value of the corresponding variable at the time step t_k. It can be mentioned here that the MPSP method uses a predictor–corrector approach. In the prediction step, a more accurate numerical integration technique (such as forth order Runge–Kutta method [7]) can be used, whereas the control sequence can still be computed using the discretized system dynamics arrived from the Euler integration formula. Such a 'hybrid approach' usually leads to faster convergence with reduced number of iterations [6].

The objective of the satellite formation flying problem is to guide the deputy satellite in such a manner so as to achieve a desired relative separation at the end of the guidance phase. Moreover, the velocity components should also match with the corresponding desired orbital parameters. Mathematically, this leads to the following two conditions at $t = t_f$: (i) $\left[x_1 \; x_3 \; x_5 \right]^T \to \left[x_1^* \; x_3^* \; x_5^* \right]^T$ and (ii) $\left[x_2 \; x_4 \; x_6 \right]^T \to \left[x_2^* \; x_4^* \; x_6^* \right]^T$. Because of this observation, the system output at the final time (N) can be defined as

$$\mathbf{y}_N \triangleq \mathbf{x}_N \tag{7.41}$$

Then, the objective of the guidance is that the following condition should be achieved at $k = N$.

$$\mathbf{y}_N = \mathbf{y}_N^* \tag{7.42}$$

where $\mathbf{y}_N^*$ is the desired output vector. The same objective can be rephrased as

$$d\mathbf{y}_N \triangleq (\mathbf{y}_N^* - \mathbf{y}_N) = 0 \tag{7.43}$$

Next, the aim is to compute the control command sequence $\mathbf{u}_k, k = 1, \ldots, N-1$ such that $d\mathbf{y_N} \to 0$. To achieve this objective, following the discrete MPSP theory outlined in Section 7.1, the coefficients $B_1, \ldots, B_{N-1}$ are evaluated using Eqs. (7.25)–(7.27).

The partial derivative of $F(\mathbf{x}_k, \mathbf{u}_k)$ and $\mathbf{y}_N$ required to compute the sensitive matrices B_k is computed as follows:

$$\frac{\partial F(\mathbf{x}_k, \mathbf{u}_k)}{\partial \mathbf{x}_k} = I_6 + \Delta t \left[\frac{\partial f_k}{\partial \mathbf{x}_k}\right] \tag{7.44}$$

$$\frac{\partial F(\mathbf{x}_k, \mathbf{u}_k)}{\partial \mathbf{u}_k} = \Delta t \begin{bmatrix} 0\,1\,0\,0\,0\,0 \\ 0\,0\,0\,1\,0\,0 \\ 0\,0\,0\,0\,0\,1 \end{bmatrix}^T \tag{7.45}$$

$$\frac{\partial \mathbf{y}_N}{\partial \mathbf{x}_N} = I_6 \tag{7.46}$$

The component of the partial derivative term $\left[\frac{\partial f_k}{\partial \mathbf{x}_k}\right]$ at every time step k is derived from the following expressions

$$\frac{\partial f_1}{\partial x_{2_k}} = 1$$

$$\frac{\partial f_2}{\partial x_{1_k}} = \dot{\nu}^2 - \mu \left[\frac{\gamma_k - 3x_{1_k}\gamma_k^{\frac{1}{2}}\left(r_c + x_{1_k}\right)}{\gamma_k^2}\right] + 3\mu r_c \gamma_k^{-\frac{5}{2}}\left(r_c + x_{1_k}\right)$$

$$\frac{\partial f_2}{\partial x_{3_k}} = \ddot{\nu} + 3\mu\gamma_k^{-\frac{5}{2}} x_{3_k}\left(r_c + x_{1_k}\right)$$

$$\frac{\partial f_2}{\partial x_{4k}} = 2\dot{\nu}$$

$$\frac{\partial f_2}{\partial x_{5_k}} = 3\mu\left(r_c + x_{1_k}\right)\gamma_k^{-\frac{5}{2}} x_{5_k}$$

$$\frac{\partial f_3}{\partial x_{4_k}} = 1$$

$$\frac{\partial f_4}{\partial x_{1_k}} = -\ddot{\nu} - 3\mu x_{3_k}\gamma_k^{-\frac{5}{2}}\left(r_c + x_{1_k}\right)$$

$$\frac{\partial f_4}{\partial x_{2_k}} = -2\dot{\nu}$$

$$\frac{\partial f_4}{\partial x_{3_k}} = \dot{\nu}^2 + \mu\left[\frac{\gamma_k - 3x_{3_k}^2\gamma_k^{\frac{1}{2}}}{\gamma_k^2}\right]$$

$$\frac{\partial f_4}{\partial x_{5_k}} = -3\mu x_{3_k} x_{5_k}\gamma_k^{-\frac{5}{2}}$$

$$\frac{\partial f_5}{\partial x_{6_k}} = 1$$

$$\frac{\partial f_6}{\partial x_{1_k}} = 3\mu x_{5_k}\left(r_c + x_{1_k}\right)\gamma_k^{-\frac{5}{2}}$$

$$\frac{\partial f_6}{\partial x_{3_k}} = 3\mu x_{3_k} x_{5_k} \gamma_k^{-\frac{5}{2}}$$

$$\frac{\partial f_6}{\partial x_{5_k}} = -\mu \left[\frac{\gamma_k - 3x_{5_k}\gamma_k^{\frac{1}{2}}}{\gamma_k^2} \right]$$

The partial derivative of x-component of J_2 gravitational harmonics $a_{J_{2x}}$ is derived (refer to Eqs. (2.77) and (2.78)) from the following expression

$$\frac{\partial a_{J_{2x}}}{\partial x_{1_k}} = -\frac{3}{2}\mu J_2 R_e^2 \Big[\frac{4x_{1_k}}{(r_c+\rho_k)^5\sqrt{\rho_k}} - \frac{4x_{1_k}}{(r_c+\rho_k)^5\sqrt{\rho_k}} (3\sin^2(i_c+\delta i_k)\sin^2(\Theta_c+\delta\Theta_k)) + \frac{3}{(r_c+\rho_k)^4} \Big\{ 2\sin(i_c+\delta i_k)\cos(i_c+\delta i_k)\,\Sigma_{31}^{-1}\sin^2(\Theta_c+\delta\Theta_k) + 2\sin^2(i_c+\delta i_k) \sin(\Theta_c+\delta\Theta_k)\cos(\Theta_c+\delta\Theta_k)\,\Sigma_{21}^{-1} \Big\} \Big] \tag{7.47}$$

$$\frac{\partial a_{J_{2x}}}{\partial x_{2_k}} = -\frac{3}{2}\mu J_2 R_e^2 \Big[\frac{3}{(r_c+\rho_k)^4} \Big\{ 2\sin(i_c+\delta i_k)\cos(i_c+\delta i_k)\,\Sigma_{32}^{-1}\sin^2(\Theta_c+\delta\Theta_k) + 2\sin^2(i_c+\delta i_k)\sin(\Theta_c+\delta\Theta_k)\cos(\Theta_c+\delta\Theta_k)\,\Sigma_{22}^{-1} \Big\} \Big] \tag{7.48}$$

$$\frac{\partial a_{J_{2x}}}{\partial x_{3_k}} = -\frac{3}{2}\mu J_2 R_e^2 \Big[\frac{4x_{3_k}}{(r_c+\rho_k)^5\sqrt{\rho_k}} - \frac{4x_{3_k}}{(r_c+\rho_k)^5\sqrt{\rho_k}} (3\sin^2(i_c+\delta i_k)\sin^2(\Theta_c+\delta\Theta_k)) + \frac{3}{(r_c+\rho_k)^4} \Big\{ 2\sin(i_c+\delta i_k)\cos(i_c+\delta i_k)\,\Sigma_{33}^{-1}\sin^2(\Theta_c+\delta\Theta_k) + 2\sin^2(i_c+\delta i_k) \sin(\Theta_c+\delta\Theta_k)\cos(\Theta_c+\delta\Theta_k)\,\Sigma_{23}^{-1} \Big\} \Big] \tag{7.49}$$

$$\frac{\partial a_{J_{2x}}}{\partial x_{4_k}} = -\frac{3}{2}\mu J_2 R_e^2 \Big[\frac{3}{(r_c+\rho_k)^4} \Big\{ 2\sin(i_c+\delta i_k)\cos(i_c+\delta i_k)\,\Sigma_{34}^{-1}\sin^2(\Theta_c+\delta\Theta_k) + 2\sin^2(i_c+\delta i_k)\sin(\Theta_c+\delta\Theta_k)\cos(\Theta_c+\delta\Theta_k)\,\Sigma_{24}^{-1} \Big\} \Big] \tag{7.50}$$

$$\frac{\partial a_{J_{2x}}}{\partial x_{5_k}} = -\frac{3}{2}\mu J_2 R_e^2 \Big[\frac{4x_{5_k}}{(r_c+\rho_k)^5\sqrt{\rho_k}} - \frac{4x_{5_k}}{(r_c+\rho_k)^5\sqrt{\rho_k}} (3\sin^2(i_c+\delta i_k)\sin^2(\Theta_c+\delta\Theta_k)) + \frac{3}{(r_c+\rho_k)^4} \Big\{ 2\sin(i_c+\delta i_k)\cos(i_c+\delta i_k)\,\Sigma_{35}^{-1}\sin^2(\Theta_c+\delta\Theta_k) + 2\sin^2(i_c+\delta i_k) \sin(\Theta_c+\delta\Theta_k)\cos(\Theta_c+\delta\Theta_k)\,\Sigma_{25}^{-1} \Big\} \Big] \tag{7.51}$$

$$\frac{\partial a_{J_{2x}}}{\partial x_{6_k}} = -\frac{3}{2}\mu J_2 R_e^2 \Big[\frac{3}{(r_c+\rho_k)^4} \Big\{ 2\sin(i_c+\delta i_k)\cos(i_c+\delta i_k)\,\Sigma_{34}^{-1}\sin^2(\Theta_c+\delta\Theta_k) + 2\sin^2(i_c+\delta i_k)\sin(\Theta_c+\delta\Theta_k)\cos(\Theta_c+\delta\Theta_k)\,\Sigma_{24}^{-1} \Big\} \Big] \tag{7.52}$$

Note that the remaining partial derivative terms turn out to be zeros. All symbols and constants are explained properly in Chapter 2. The partial derivatives of y, z components of J_2 gravitational harmonics can be evaluated in a similar manner. The expressions are fairly similar, and hence details are omitted for brevity.

The cost function is selected such that the control effort is minimized which is represented by

$$J = \frac{1}{2}\sum_{k=1}^{N-1} \mathbf{u}_k{}^T R_k \mathbf{u}_k \tag{7.53}$$

where $\mathbf{u}_k = \begin{bmatrix} a_{x_k} & a_{y_k} & a_{z_k} \end{bmatrix}^T$ are the applied control accelerations on the deputy satellite.

After the sensitivity matrix is calculated, A_{λ^i} and b_{λ^i} are evaluated using Eq. (7.35). Finally, the improvised control at every step $\mathbf{u}_k$ is computed from Eq. (7.38). This iterative process is repeated as shown in Fig. 7.1 until the final output vector $\mathbf{y}_N$ achieves the desired value $\mathbf{y}_N^*$.

Note that there is the requirement of a reasonably good guess history to start the MPSP algorithm. The F-LQR solution approach discussed in Chapter 6 can be used for this purpose. It is to be noted that if the reader is not interested in J_2 gravitational harmonics, then all the corresponding terms relating to it has to be assumed zero for the analysis.

7.3 Discrete MPSP: Results and Discussions

The set of results included here consist of three cases: (i) circular orbit of the chief satellite and small desired relative separation (mild case), (ii) elliptic orbit of the chief satellite with small eccentricity and small desired relative separation (medium complexity), and (iii) elliptic orbit of the chief satellite with large eccentricity and large desired relative separation and that too in presence of J_2 gravitational harmonics (severe case). The spectrum of scenarios demonstrates the wide and generic applicability of the MPSP guidance. In case (ii), the results are also compared with the finite-time SDRE (F-SDRE) formulation to show that the MPSP approach leads to better terminal accuracy. Note that F-SDRE which is based on finite-time formulation makes it compatible to the MPSP formulation. Case (iii) demonstrates the wider applicability of the MPSP formulation in presence of complex nonlinear terms in the system dynamics. Note that, without loss of generality, results from multiple initial conditions are included only for case (i).

Simulation results provided in this section can be obtained from the program files provided in the folder named '*MPSP*'. For more details, refer to Section A.5.

7.3.1 Formation Flying in Circular Orbit and with Small Desired Relative Distance

The initial condition and the desired final condition for this simulation study are selected as per Table 2.2 in Chapter 2.

Based on MPSP guidance solution, Fig. 7.2 shows the trajectories from different initial angle $\phi_i = 30°, 150°, 245°$ to the corresponding desired final angle $\phi_f = 45°, 165°, 260°$ of deputy satellite position vector with respect to chief satellite velocity vector (refer to Fig. 2.8). It is clearly seen that for every initial condition, the MPSP guidance performs satisfactorily and, as expected, is able to drive the deputy satellite to the corresponding desired final condition.

7.3.2 Formation Flying in Elliptic Orbit and with Small Desired Relative Distance

The initial condition and the desired final conditions for this simulation study are selected as per Table 2.3 in Chapter 2.

The trajectory of the deputy satellite in Hill's frame is shown in Fig. 7.3. The relative position and velocity error histories are plotted in Figs. 7.4 and 7.5, respectively. The achieved error values in the states at the final time are tabulated in Table 7.1. The necessary control history (i.e., the guidance command history) is plotted in Fig. 7.6.

It can be observed that the deputy satellite is able to achieve the desired relative formation using both MPSP and F-SDRE approaches. However, the terminal error based on the F-SDRE approach is relatively higher as compared to the state error based on MPSP, since the MPSP approach exploits the nonlinear plant model much better, whereas the F-SDRE approach relies on linear optimal control

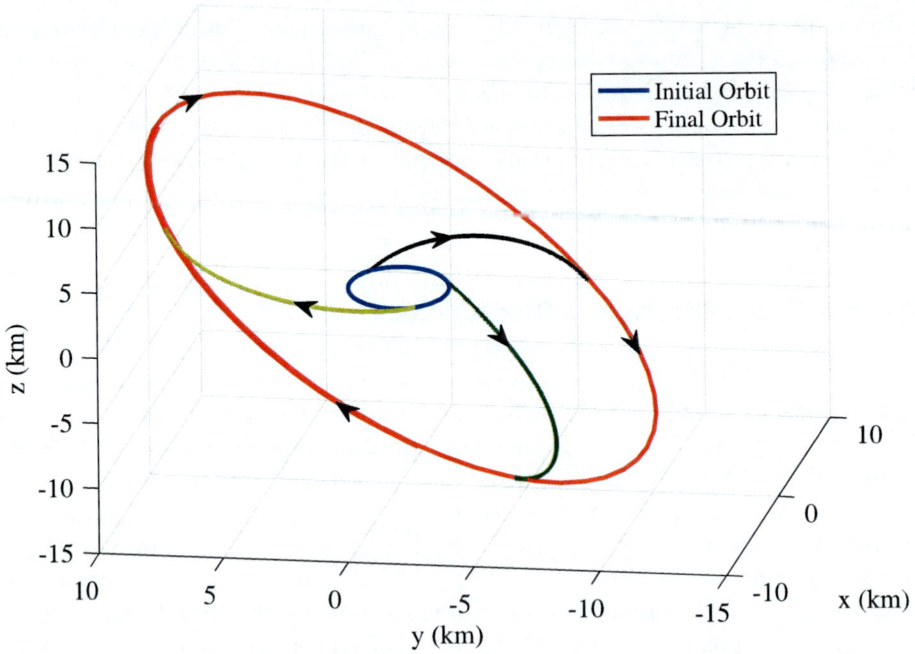

Fig. 7.2 Deputy satellite (DS) trajectories ($\rho_i = 1\,\text{km}$, $\rho_f = 5\,\text{km}$, $e = 0.0$)

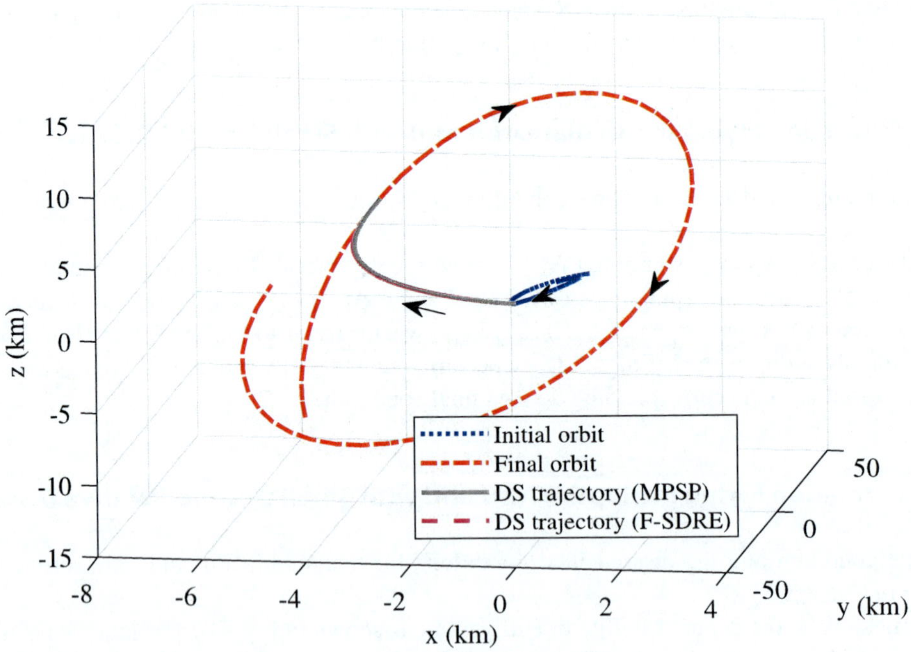

Fig. 7.3 Deputy satellite (DS) trajectory ($\rho_i = 1\,\text{km}$, $\rho_f = 5\,\text{km}$, $e = 0.05$)

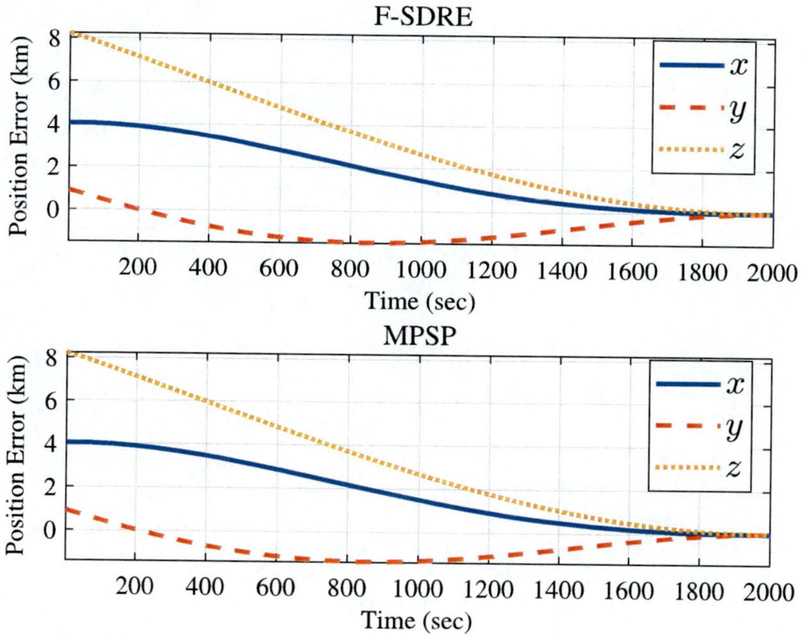

Fig. 7.4 Position error history ($\rho_i = 1$ km, $\rho_f = 5$ km, $e = 0.05$)

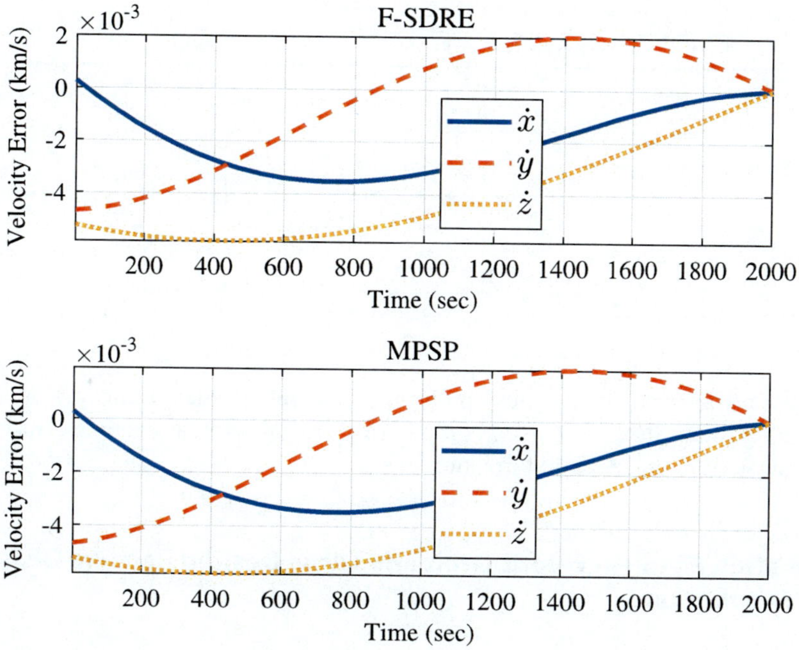

Fig. 7.5 Velocity error history ($\rho_i = 1$ km, $\rho_f = 5$ km, $e = 0.05$)

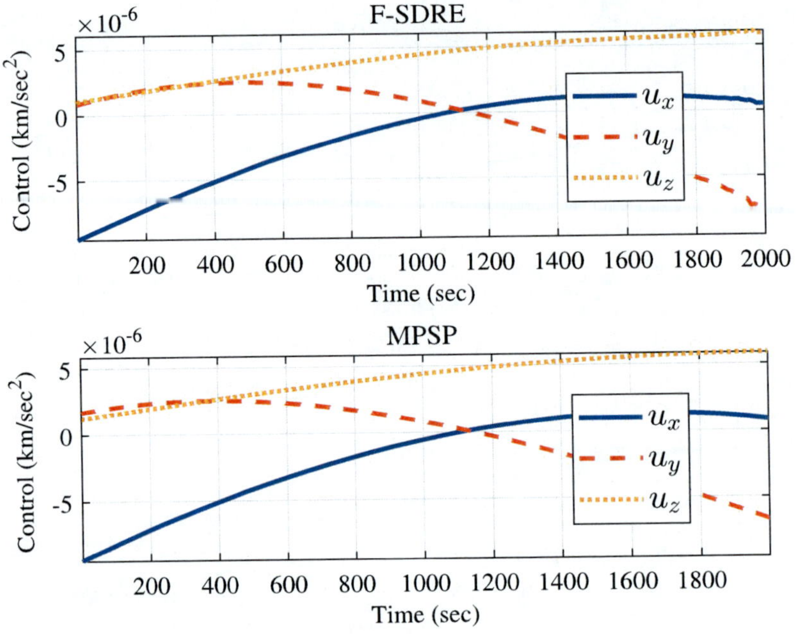

Fig. 7.6 Control history for deputy satellite ($\rho_i = 1$ km, $\rho_f = 5$ km, $e = 0.05$)

Table 7.1 Terminal state error ($\rho_i = 1$ km, $\rho_f = 5$ km, $e = 0.05$)

State error	F-SDRE	MPSP
x (km)	1.75×10^{-3}	-2.59×10^{-5}
$\dot{x}$ (km/s)	5.29×10^{-6}	-8.29×10^{-8}
y (km)	3.78×10^{-3}	-5.61×10^{-5}
$\dot{y}$ (km/s)	-4.95×10^{-5}	-1.45×10^{-7}
z (km)	7.34×10^{-3}	-5.72×10^{-10}
$\dot{z}$ (km/s)	1.88×10^{-5}	-3.37×10^{-12}
Norm	8.44×10^{-3}	6.17×10^{-5}

theory to solve a nonlinear optimal control problem. For a SFF problem achieving the final velocity states along with position states on the final orbit is very crucial. Else, as the time elapses, the deputy satellite drifts away from the required formation.

7.3.3 Formation Flying in Elliptic Orbit and Large Desired Relative Distance with J_2 Perturbation

Here, the chief satellite is assumed to fly in the elliptic orbit having relatively large eccentricity of 0.2 and large desired relative distance (refer to Table 2.6 in Chapter 2). Moreover, the J_2 gravitational harmonics is also included in the problem formulation.

The resulting relative trajectory of deputy satellite in Hill's frame is shown in Fig. 7.7. The deputy satellite starts from the inner initial relative formation trajectory and is commanded to the outer relative orbit. As shown in Fig. 7.7, the MPSP guidance is capable of driving the deputy satellite to the desired

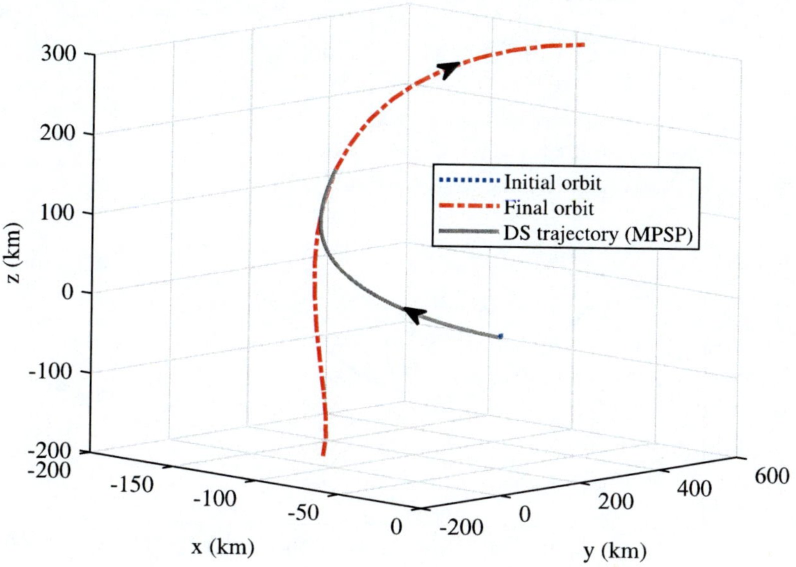

Fig. 7.7 Deputy satellite (DS) trajectory ($\rho_i = 1\,\text{km}$, $\rho_f = 100\,\text{km}$, $e = 0.2$)

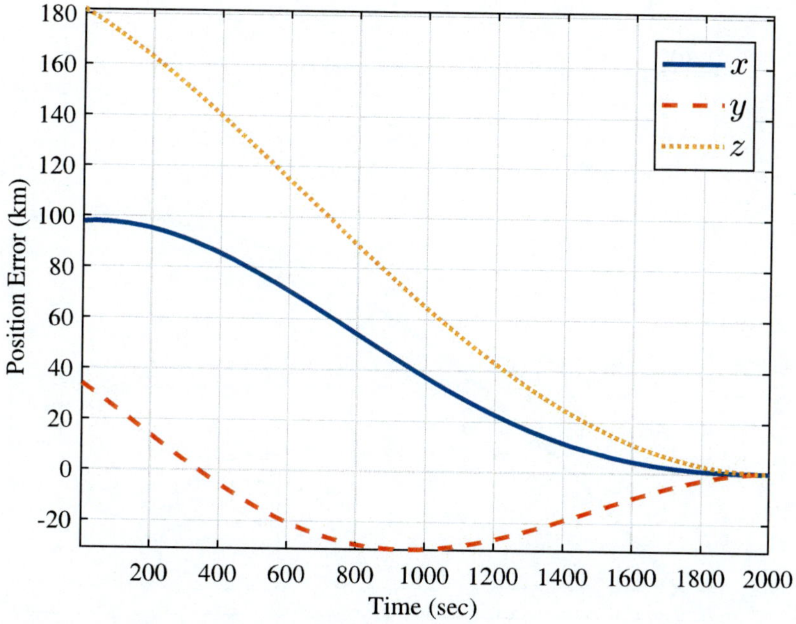

Fig. 7.8 Position error history ($\rho_i = 1\,\text{km}$, $\rho_f = 100\,\text{km}$, $e = 0.2$)

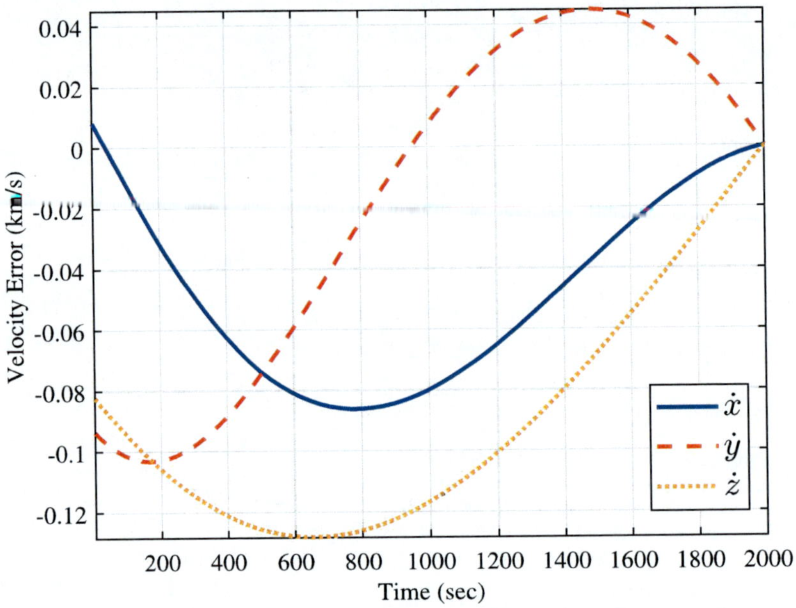

Fig. 7.9 Velocity error history ($\rho_i = 1$ km, $\rho_f = 100$ km, $e = 0.2$)

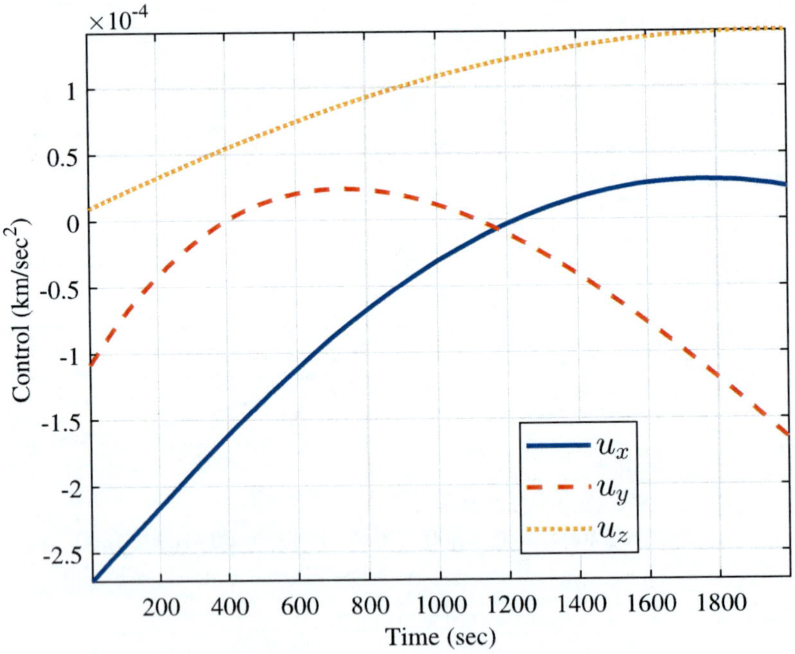

Fig. 7.10 Control history for deputy satellite ($\rho_i = 1$ km, $\rho_f = 100$ km, $e = 0.2$)

relative states even though the chief satellite is in the high eccentric orbit and the relative distance between the deputy and chief satellite is large. The state error history is plotted in Figs. 7.8 and 7.9 which shows the state errors reduce along the trajectory to achieve the desired condition. Figure 7.10 illustrates the optimal control required in achieving the desired state values. The control trajectories are quite smooth, which is a desirable feature.

7.4 Generalized (Continuous) Model Predictive Static Programming: Generic Theory

Similar to MPSP approach discussed in Section 7.1, the generalized (continuous) model predictive static programming (G-MPSP) is also meant for rapidly solving a class of finite-horizon nonlinear optimal control problems with hard terminal constraints. It retains all the advantages of MPSP design. However, as an added advantage, the entire problem formulation and solution approach is in the *continuous-time* framework, and hence it does not demand discretized system dynamics and cost function to begin with. Because of this reason, any higher order numerical scheme can easily be incorporated into the solution process. The theoretical details of the generalized model predictive static programming is presented first, followed by the associated results in connection with the satellite formation flying.

For the G-MPSP design, let us consider the nonlinear system defined in Section 7.1. In continuous-time setting at the ith iteration, Eq. (7.1) results in following equations

$$\dot{\mathbf{x}}^i(t) = f\left(\mathbf{x}^i(t), \mathbf{u}^i(t)\right) \tag{7.54}$$

$$\mathbf{y}^i(t) = h\left(\mathbf{x}^i(t)\right) \tag{7.55}$$

where $\mathbf{x}^i \in \Re^n$, $\mathbf{u}^i \in \Re^m$, $\mathbf{y}^i \in \Re^p$ are the state, control and output vectors respectively. The primary objective is to obtain a suitable control history $\mathbf{u}^i(t)$ so that the output $\mathbf{y}^i(t_f)$ at the fixed final time t_f goes to a desired value $\mathbf{y}^*(t_f)$, i.e., $\mathbf{y}^i(t_f) \rightarrow \mathbf{y}^*(t_f)$, and this should be achieved with minimum control effort.

Similar to MPSP technique, the G-MPSP approach needs to start from a 'guess history' of the guidance solution. Based on a continuous domain approach, a way to compute an error history of the control variable is presented, which needs to be added to the guess history to get an improved guidance history. This iteration continues until the objective is met, i.e., until $\mathbf{y}^i(t_f) \rightarrow \mathbf{y}^*(t_f)$. Note that the G-MPSP technique also produces control history update in a closed form, thereby making it computationally very efficient. Derivation of the entire algorithm is presented in the following paragraphs.

Multiplying both sides of Eq. (7.54) by a matrix $W^i(t)$ produces

$$W^i(t)\,\dot{\mathbf{x}}^i = W^i(t)\, f\left(\mathbf{x}^i(t), \mathbf{u}^i(t)\right) \tag{7.56}$$

The computation of the matrix $W^i(t) \in \Re^{p\times n}$ is presented in the later part of this section. One can also notice that $W^i(t)$ plays the role of projecting the system dynamics to the output space (which is typically of lesser dimension as compared to the state-space dimension). This reduced dynamics is considered in further development of the G-MPSP formulation, which is a good reason for its computational efficiency.

Now, by integrating both sides of Eq. (7.56) from t_0 to t_f, one obtains

$$\int_{t_0}^{t_f}\left[W^i(t)\,\dot{\mathbf{x}}^i(t)\right]dt=\int_{t_0}^{t_f}\left[W^i(t)\,f\left(\mathbf{x}^i(t),\mathbf{u}^i(t)\right)\right]dt \tag{7.57}$$

Next, taking the left-hand side expression to the right-hand side and adding the quantity $\mathbf{y}^i\left(\mathbf{x}^i\left(t_f\right)\right)$ to both sides, the following equation is obtained.

$$\mathbf{y}^i\left(\mathbf{x}^i\left(t_f\right)\right)=\mathbf{y}^i\left(\mathbf{x}^i\left(t_f\right)\right)+\int_{t_0}^{t_f}\left[W^i(t)\,f\left(\mathbf{x}^i(t),\mathbf{u}^i(t)\right)\right]dt-\int_{t_0}^{t_f}\left[W^i(t)\,\dot{\mathbf{x}}^i(t)\right]dt \tag{7.58}$$

Now, consider only the last term of the Eq. (7.58) and perform integration by parts to obtain

$$\begin{aligned}\int_{t_0}^{t_f}\left[W^i(t)\,\dot{\mathbf{x}}^i(t)\right]dt&=\left[W^i(t)\,\mathbf{x}^i(t)\right]_{t_0}^{t_f}-\int_{t_0}^{t_f}\left[\dot{W}^i(t)\,\mathbf{x}^i(t)\right]dt\\&=\left[W^i\left(t_f\right)\mathbf{x}^i\left(t_f\right)-W^i\left(t_0\right)\mathbf{x}^i\left(t_0\right)\right]-\int_{t_0}^{t_f}\left[\dot{W}^i(t)\,\mathbf{x}^i(t)\right]dt\end{aligned} \tag{7.59}$$

Substituting Eq. (7.59) in Eq. (7.58) leads to the following equation

$$\begin{aligned}\mathbf{y}^i\left(\mathbf{x}^i\left(t_f\right)\right)&=\mathbf{y}^i\left(\mathbf{x}^i\left(t_f\right)\right)-\left[W^i\left(t_f\right)\mathbf{x}^i\left(t_f\right)\right]+\left[W^i\left(t_0\right)\mathbf{x}^i\left(t_0\right)\right]\\&\quad+\int_{t_0}^{t_f}\left[W^i(t)\,f\left(\mathbf{x}^i(t),\mathbf{u}^i(t)\right)+\dot{W}^i(t)\,\mathbf{x}^i(t)\right]dt\end{aligned} \tag{7.60}$$

Performing, the first variational operation on both sides of Eq. (7.60) results in

$$\begin{aligned}\delta\mathbf{y}^i\left(\mathbf{x}^i\left(t_f\right)\right)&=\left[\frac{\partial\mathbf{y}^i\left(\mathbf{x}^i(t)\right)}{\partial\mathbf{x}^i(t)}\delta\mathbf{x}^i(t)\right]_{t=t_f}-W^i\left(t_f\right)\delta\mathbf{x}^i\left(t_f\right)+W^i\left(t_0\right)\delta\mathbf{x}^i\left(t_0\right)\\&\quad+\int_{t_0}^{t_f}\left[\left(W^i(t)\,\frac{\partial f\left(\mathbf{x}^i(t),\mathbf{u}^i(t)\right)}{\partial\mathbf{x}^i(t)}\right)\delta\mathbf{x}^i(t)\right.\\&\quad\left.+\left(W^i(t)\,\frac{\partial f\left(\mathbf{x}^i(t),\mathbf{u}^i(t)\right)}{\partial\mathbf{u}^i(t)}\right)\delta\mathbf{u}^i(t)+\dot{W}^i(t)\,\delta\mathbf{x}^i(t)\right]dt\\&=\left[\left(\frac{\partial\mathbf{y}^i\left(\mathbf{x}^i(t)\right)}{\partial\mathbf{x}^i(t)}-W^i(t)\right)\delta\mathbf{x}^i(t)\right]_{t=t_f}+W^i\left(t_0\right)\delta\mathbf{x}^i\left(t_0\right)\\&\quad+\int_{t_0}^{t_f}\left[\left(W^i(t)\,\frac{\partial f\left(\mathbf{x}^i(t),\mathbf{u}^i(t)\right)}{\partial\mathbf{x}^i(t)}+\dot{W}^i(t)\right)\delta\mathbf{x}^i(t)\right.\\&\quad\left.+\left(W^i(t)\,\frac{\partial f\left(\mathbf{x}^i(t),\mathbf{u}^i(t)\right)}{\partial\mathbf{u}^i(t)}\right)\delta\mathbf{u}^i(t)\right]dt\end{aligned} \tag{7.61}$$

Next, it is desired to determine the variation $\delta\mathbf{y}^i\left(\mathbf{x}^i\left(t_f\right)\right)$ produced by the variation $\delta\mathbf{u}^i(t)$ alone. This can be done by choosing $\dot{W}^i(t)$ in a way such that it causes the coefficients of $\delta\mathbf{x}^i(t)$ in the Eq. (7.61) to vanish. This results in the following equations

$$\dot{W}^i(t) = -W^i(t)\left(\frac{\partial f\left(\mathbf{x}^i(t), \mathbf{u}^i(t)\right)}{\partial \mathbf{x}^i(t)}\right) \tag{7.62}$$

$$W^i\left(t_f\right) = \frac{\partial \mathbf{y}^i\left(\mathbf{x}^i\left(t_f\right)\right)}{\partial \mathbf{x}^i\left(t_f\right)} \tag{7.63}$$

Note that Eq. (7.62) can be integrated backward starting from the boundary condition available in Eq. (7.63). Moreover, there is no variation (error) in the initial condition as it is specified (typically the case in all practical systems). Hence,

$$\delta \mathbf{x}^i(t_0) = 0 \tag{7.64}$$

With these observations, substituting the results in Eqs. (7.62), (7.63) and (7.64), Eq. (7.61) can be simplified as

$$\delta \mathbf{y}^i\left(\mathbf{x}^i\left(t_f\right)\right) = \int_{t_0}^{t_f}\left[B_c^i(t)\,\delta \mathbf{u}^i(t)\right]dt \tag{7.65}$$

where

$$B_c^i(t) = W^i(t)\,\frac{\partial f\left(\mathbf{x}^i(t), \mathbf{u}^i(t)\right)}{\partial \mathbf{u}^i(t)} \tag{7.66}$$

Note that Eqs. (7.62), (7.63), (7.66) are direct correlation of the recursive computation of sensitivity matrix of discrete MPSP (see Eqs. (7.25), (7.26), (7.27)). It should be mentioned here that the error in output at the final time t_f is defined as

$$\Delta \mathbf{y}^i\left(\mathbf{x}^i\left(t_f\right)\right) = \mathbf{y}^*\left(t_f\right) - \mathbf{y}^i\left(t_f\right) \tag{7.67}$$

Under small error approximations, we can write $\Delta \mathbf{y}^i\left(\mathbf{x}^i\left(t_f\right)\right) = \delta \mathbf{y}^i\left(\mathbf{x}^i\left(t_f\right)\right)$. Hence, $\delta \mathbf{y}^i\left(\mathbf{x}^i\left(t_f\right)\right)$ in Eq. (7.65) is computed utilizing Eq. (7.67).

One can notice here that Eq. (7.65) relates to the error of output at the final time t_f to the error of control history $\mathbf{u}^i(t)$ for all $t \in [t_0, t_f)$. Obviously, it is an under-constrained problem, and there is a scope for optimizing the control history. Taking advantage of this, the following performance index is considered for control effort minimization,

$$J = \frac{1}{2}\int_{t_0}^{t_f}\left[\left(\mathbf{u}^i(t) + \delta \mathbf{u}^i(t)\right)^T R(t)\left(\mathbf{u}^i(t) + \delta \mathbf{u}^i(t)\right)\right]dt \tag{7.68}$$

where $\mathbf{u}^i(t)$ is the ith control history solution (assumed to be known), $R(t)$ is the positive definite weighting matrix that needs to be chosen judiciously by the control designer. The selection of this cost function is motivated by the fact that it is desired to find an l_2-norm minimizing control history, since $\mathbf{u}^i(t) + \delta \mathbf{u}^i(t)$ is the updated control value $\mathbf{u}^{i+1}(t)$.

The cost function in Eq. (7.68), which is purely a function of $\delta \mathbf{u}^i(t)$), needs to be minimized with respect to the appropriate selection of $\delta \mathbf{u}^i(t)$), subject to the isoperimetric constraint in Eq. (7.65). To solve this problem, using the optimization theory [2], the augmented cost function is given by

$$\begin{aligned}\bar{J} = &\frac{1}{2}\int_{t_0}^{t_f}\left[\left(\mathbf{u}^i(t) + \delta \mathbf{u}^i(t)\right)^T R(t)\left(\mathbf{u}^i(t) + \delta \mathbf{u}^i(t)\right)\right]dt \\ &+ \boldsymbol{\lambda}^{i^T}\left[\delta \mathbf{y}^i\left(t_f\right) - \int_{t_0}^{t_f}\left[B_c^i(t)\,\delta \mathbf{u}^i(t)\right]dt\right]\end{aligned} \tag{7.69}$$

where $\boldsymbol{\lambda}^i$ is the Lagrange multiplier in the ith iteration. Since $\bar{J}$ is a function of free variables $\delta\mathbf{u}^i$ and $\boldsymbol{\lambda}^i$, the first variation of Eq. (7.69) is given by

$$\delta\bar{J} = \frac{\partial\bar{J}}{\partial\left(\delta\mathbf{u}^i\right)}\delta(\delta\mathbf{u}^i) + \frac{\partial\bar{J}}{\partial\boldsymbol{\lambda}^i}\delta\boldsymbol{\lambda}^i \tag{7.70}$$

It is to be noted that variation of integration is equal to the integration of variation, i.e., these operators satisfy the commutation property [8]. Using Eq. (7.69), the components of Eq. (7.70) are rewritten as follows

$$\frac{\partial\bar{J}}{\partial\left(\delta\mathbf{u}^i\right)}\delta(\delta\mathbf{u}^i) = \frac{1}{2}\int_{t_0}^{t_f}\left[\frac{\partial\left[\left(\mathbf{u}^i(t)+\delta\mathbf{u}^i(t)\right)^T R(t)\left(\mathbf{u}^i(t)+\delta\mathbf{u}^i(t)\right)\right]}{\partial(\delta\mathbf{u}^i)}\delta(\delta\mathbf{u}^i)\right]dt$$
$$-\int_{t_0}^{t_f}\left[\frac{\partial\left[\boldsymbol{\lambda}^{iT}(B_c^i(t)\delta\mathbf{u}^i(t))\right]}{\partial(\delta\mathbf{u}^i)}\delta(\delta\mathbf{u}^i)\right]dt \tag{7.71}$$
$$= \frac{1}{2}\int_{t_0}^{t_f}\left[R(t)\left(\mathbf{u}^i(t)+\delta\mathbf{u}^i(t)\right)\delta(\delta\mathbf{u}^i)\right]dt$$
$$-\int_{t_0}^{t_f}\left[B_c^T(t)\boldsymbol{\lambda}^i\delta(\delta\mathbf{u}^i)\right]dt \tag{7.72}$$
$$= \frac{1}{2}\int_{t_0}^{t_f}\left[R(t)\left(\mathbf{u}^i(t)+\delta\mathbf{u}^i(t)\right) - B_c^T(t)\boldsymbol{\lambda}^i\right]\delta(\delta\mathbf{u}^i)dt \tag{7.73}$$
$$\frac{\partial\bar{J}}{\partial\boldsymbol{\lambda}^i}\delta\boldsymbol{\lambda}^i = \frac{\partial\left[\boldsymbol{\lambda}^{iT}\left(\delta\mathbf{y}^i\left(t_f\right) - \int_{t_0}^{t_f}\left[B_c^i(t)\,\delta\mathbf{u}^i(t)\right]dt\right)\right]}{\partial\boldsymbol{\lambda}^i}\delta\boldsymbol{\lambda}^i$$
$$= \left[\delta\mathbf{y}^i\left(\mathbf{x}^i\left(t_f\right)\right) - \int_{t_0}^{t_f}\left[B_c^i(t)\,\delta\mathbf{u}^i(t)\right]dt\right]\delta\boldsymbol{\lambda}^i \tag{7.74}$$

The necessary condition of optimality is given by $\delta\bar{J} = 0$. In order to satisfy the necessary condition, all possible variations $\delta(\delta\mathbf{u}^i)$, $\delta\boldsymbol{\lambda}^i$ of Eq. (7.70) have to be zero which is ensured by the equating the integrand of the Eqs. (7.73) and (7.74) to zero, i.e.,

$$R(t)\left(\mathbf{u}^i(t)+\delta\mathbf{u}^i(t)\right) - B_c^T(t)\boldsymbol{\lambda}^i = 0 \tag{7.75}$$
$$\delta\mathbf{y}^i\left(\mathbf{x}^i\left(t_f\right)\right) - \int_{t_0}^{t_f}\left[B_c^i(t)\,\delta\mathbf{u}^i(t)\right]dt = 0 \tag{7.76}$$

Simplifying Eq. (7.75), for $\delta\mathbf{u}^i$ results in

$$\delta\mathbf{u}^i(t) = [R(t)]^{-1}\left[B_c^i(t)\right]^T\boldsymbol{\lambda}^i - \mathbf{u}^i(t) \tag{7.77}$$

Substituting Eq. (7.77) into Eq. (7.76) leads to

$$\delta\mathbf{y}^i\left(t_f\right) = \int_{t_0}^{t_f} B_c^i(t)\left[(R(t))^{-1}\left[B_c^i(t)\right]^T\boldsymbol{\lambda}^i - \mathbf{u}^i(t)\right]dt \tag{7.78}$$

Equation (7.78) can be written in simplified form as

$$\delta \mathbf{y}^i (t_f) = A_\lambda^i \boldsymbol{\lambda} - b_\lambda^i \tag{7.79}$$

where

$$A_\lambda^i \triangleq \int_{t_0}^{t_f} \left[B_c^i (t) [R (t)]^{-1} B_c^{i\,T} (t) \right] dt \tag{7.80}$$

and

$$b_\lambda^i \triangleq \int_{t_0}^{t_f} \left[B_c^i (t) \mathbf{u}^i (t) \right] dt \tag{7.81}$$

Assuming that A_λ^i is a non-singular matrix, solution for $\boldsymbol{\lambda}^i$ is obtained from Eq. (7.79) as

$$\boldsymbol{\lambda}^i = A_\lambda^{i\,-1} \left[\delta \mathbf{y}^i (t_f) + b_\lambda^i \right] \tag{7.82}$$

Next, the variation in control is obtained by substituting Eq. (7.82) into Eq. (7.77) which gives

$$\delta \mathbf{u}^i (t) = R (t)^{-1} B_c^i (t)^T \left[A_\lambda^{i\,-1} \left(\delta \mathbf{y}^i (t_f) + b_\lambda^i \right) \right] - \mathbf{u}^i (t) \tag{7.83}$$

Hence, the updated control is given by

$$\mathbf{u}^{i+1} (t) = \mathbf{u}^i (t) + \delta \mathbf{u}^i (t) = R (t)^{-1} B_c^i (t)^T \left[A_\lambda^{i\,-1} \left(\delta \mathbf{y}^i (t_f) + b_\lambda^i \right) \right] \tag{7.84}$$

It is clear from Eq. (7.84) that the updated control history solution is a *closed-form solution*. Furthermore, the necessary error coefficients $B_c^i(t)$ in Eq. (7.66) are computed recursively using Eq. (7.62). Overall it leads to a very fast computation of the control history update, and in turn a computationally very efficient technique.

Interestingly, the continuous-time G-MPSP technique turns out to be equivalent to the discrete-time MPSP technique when Euler integration scheme is used to propagate the system dynamics and the weighting matrix dynamics and, in addition, the rectangular rule with left grid point approximation is used to discretize the performance index. For more details on it, one can refer to [9].

7.5 G-MPSP Implementation Algorithm

The G-MPSP technique is an iterative algorithm which starts from a guess history and continues until the desired accuracy in the terminal error of the output $\mathbf{y}(t_f)$ is achieved and, simultaneously, the cost function gets converged. The following algorithmic steps are performed in every guidance cycle.

(i) Initialize the control history $\mathbf{u}^i(t)$, $i = 1$ with a 'working guess'. Theoretically, the guess trajectory should not be very far from the optimal trajectory. In our experience, however, this is not a strong requirement. In case the guess is not close to the optimal solution, it takes a few more iterations to converge.

(ii) Using the known initial condition, propagate the system dynamics in Eq. (7.54) using $\mathbf{u}^i(t)$ until t_f to get the final state of the system dynamics $\mathbf{x}^i(t_f)$ and then compute output $\mathbf{y}^i(t_f)$. Since, the required output $\mathbf{y}^*(t_f)$ is known, $\delta\mathbf{y}^i(t_f)$ is computed using Eq. (7.67).

(iii) If either of $\delta\mathbf{y}^i(t_f)$ or the cost function have not converged, then go to the next step. On the other hand, in case of convergence, use the converged control solution as the guidance command.

(iv) In case the solution has not converged, compute the update matrix $W^i(t)$ using Eq. (7.62) at each time step t using a higher order numerical integration scheme (e.g., fourth-order Runge–Kutta method).
(v) Use the matrix $W^i(t)$ to compute the $B_c^i(t)$ using Eq. (7.66).
(vi) Once $B_c^i(t)$ is computed, A_{λ^i} and b_{λ^i} are computed using Eq. (7.80) and Eq. (7.81), respectively.
(vii) Finally, compute the updated control $\mathbf{u}^{i+1}(t)$ using Eq. (7.84) and then repeat the process from step (ii).

7.6 G-MPSP: Results and Discussions

Even though MPSP and G-MPSP essentially address the same problem (one in the discrete domain and other in the continuous domain) and philosophies are more or less similar, the two techniques are independent of each other. The reader need not understand one to understand the other. Because of this reason, to demonstrate the capability of the G-MPSP guidance, the same scenarios considered in the MPSP section are considered again to maintain consistency. This is without the loss of generality in the sense that it works for other scenarios as well. The set of results included here consist of two cases: (i) elliptic orbit of the chief satellite with small eccentricity and small desired relative separation (medium complexity), and (ii) elliptic orbit of the chief satellite with large eccentricity and large desired relative separation and in the presence of J_2 gravitational harmonics (severe case). Note that the most benign case, namely, the formation flying in circular orbit with small relative distance, is not included here to limit the length of the chapter.

7.6.1 Formation Flying in Elliptic Orbit and with Small Desired Relative Distance

The initial condition and the desired final conditions for this simulation study are selected as per Table 2.3 in Chapter 2.

The trajectory of the deputy satellite in Hill's frame is shown in Fig. 7.11. The relative position and velocity error histories are plotted in Figs. 7.12 and 7.13, respectively. The necessary control history (i.e., the guidance command history) is plotted in Fig. 7.14. The achieved error values in the states at the final time are tabulated in Table 7.2. It is obvious that the deputy satellite is able to achieve the desired relative formation using both G-MPSP and finite-time SDRE (F-SDRE) method. However, the terminal error based on the F-SDRE approach is relatively much higher as compared to the state error based on G-MPSP. This is because the G-MPSP approach computes the desired control accounting for the complete nonlinear system dynamics from nonlinear optimal control theory, whereas the F-SDRE method relies only on the state-dependent coefficient form of the system dynamics, followed by linear optimal control theory.

Table 7.2 Terminal state error ($\rho_i = 1$ km, $\rho_f = 5$ km, $e = 0.05$)

State error	F-SDRE	G-MPSP
x (km)	1.75×10^{-3}	-3.10×10^{-4}
$\dot{x}$ (km/s)	5.29×10^{-6}	-7.44×10^{-7}
y (km)	3.78×10^{-3}	-4.57×10^{-4}
$\dot{y}$ (km/s)	-4.95×10^{-5}	-7.49×10^{-7}
z (km)	7.34×10^{-3}	-9.44×10^{-9}
$\dot{z}$ (km/s)	1.88×10^{-5}	-4.75×10^{-11}
Norm	8.44×10^{-3}	5.50×10^{-4}

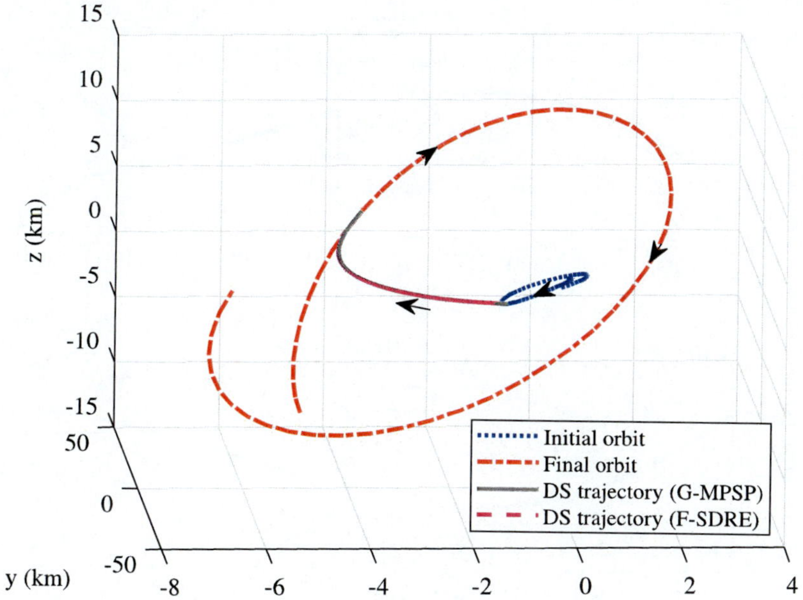

Fig. 7.11 Deputy satellite (DS) trajectory ($\rho_i = 1$ km, $\rho_f = 5$ km, $e = 0.05$)

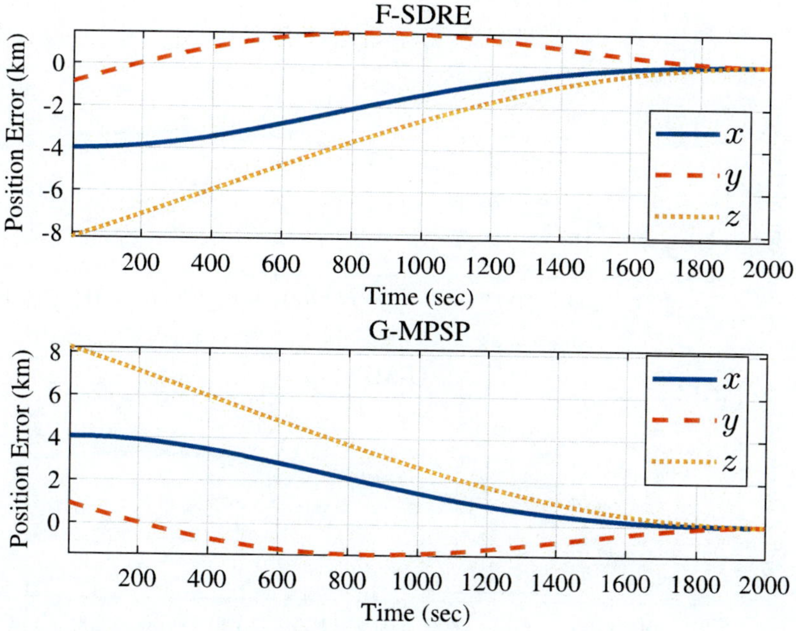

Fig. 7.12 Position error history ($\rho_i = 1$ km, $\rho_f = 5$ km, $e = 0.05$)

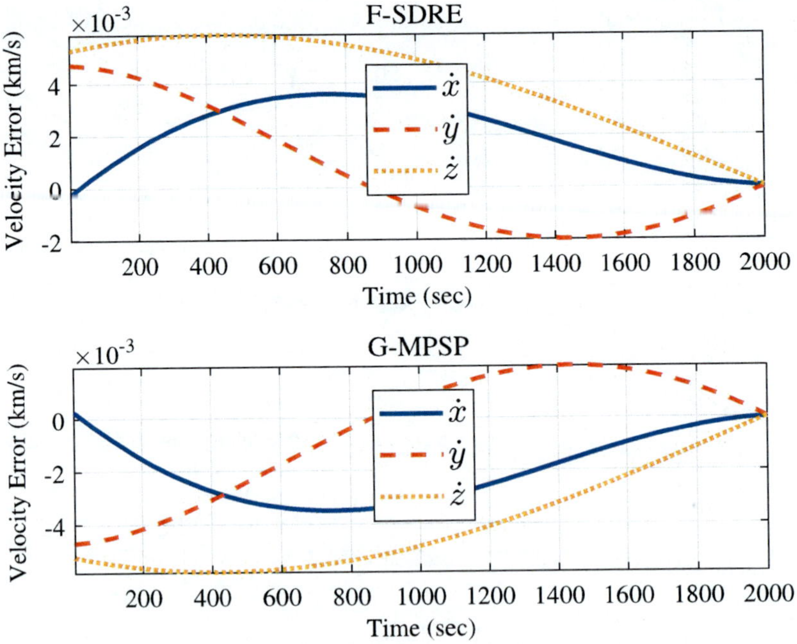

Fig. 7.13 Velocity error history ($\rho_i = 1$ km, $\rho_f = 5$ km, $e = 0.05$)

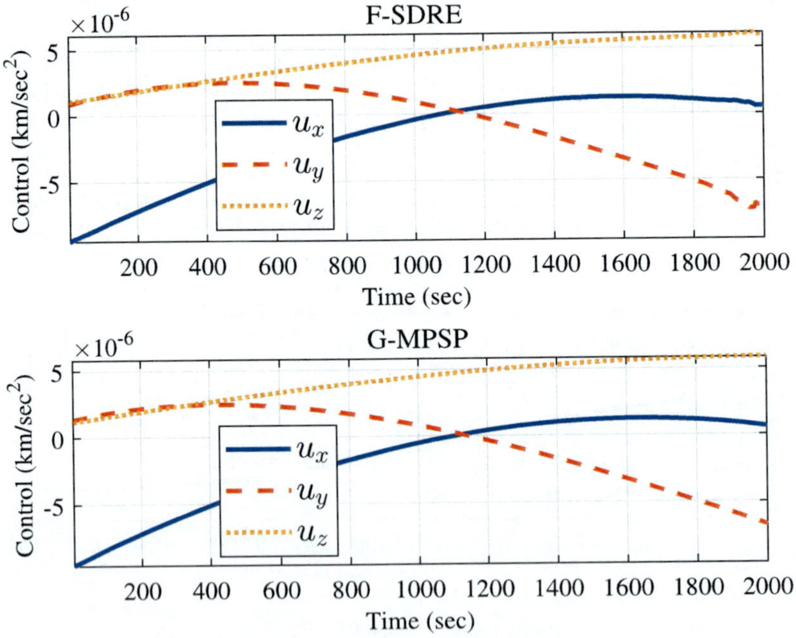

Fig. 7.14 Control history for deputy satellite ($\rho_i = 1$ km, $\rho_f = 5$ km, $e = 0.05$)

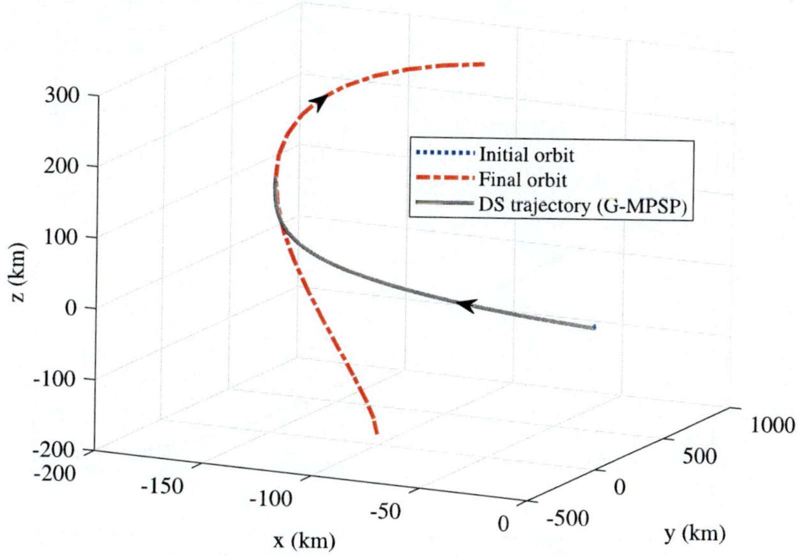

Fig. 7.15 Deputy satellite (DS) trajectory ($\rho_i = 1\,\text{km}$, $\rho_f = 100\,\text{km}$, $e = 0.2$)

7.6.2 Formation Flying in Elliptic Orbit and Large Desired Relative Distance with J_2 Perturbation

Here, the chief satellite is assumed to fly in the elliptic orbit having relatively large eccentricity of 0.2 and large desired relative distance (refer to Table 2.6 in Chapter 2). Moreover, the J_2 gravitational harmonics is also included in the problem formulation itself, i.e., it is not considered as a perturbation.

Figure 7.15 shows the trajectory from initial to final desired orbit. The relative position and velocity error history converges to zero as shown in Figs. 7.16 and 7.17, respectively. G-MPSP control solution plotted in Fig. 7.18 tries to minimize the control and achieve the final states as hard constraints even in such a penalizing case.

7.6.3 Comparison of MPSP and G-MPSP

Even though the MPSP and G-MPSP philosophies are same, earlier one is in the discrete domain, whereas the later one is in the continuous domain. Hence, naturally curiosity arises about their performance comparison. Toward this objective, the state error achieved after each iteration are tabulated in Tables 7.3 and 7.4. Note that this result is only for the formation flying in elliptic orbit and small desired relative distance scenario. The results are similar for the other case as well and hence not included here to contain the length of the chapter.

It can be observed that for achieving the desired output within the specified tolerance, G-MPSP is converging at third iteration itself, which is lesser than the number of iterations MPSP takes (four iterations). Moreover, starting from the same LQR guess solution, after the first iteration itself, G-MPSP is performing much better in achieving the desired output than MPSP. This is because higher order numerical methods are used in computing the weighting matrices. However, because of this, the G-MPSP method requires more computations per iteration. It is also true that if one uses simplified numerical procedures, G-MPSP and MPSP are same. For more details on the equivalence of G-MPSP and MPSP, one can refer to [9].

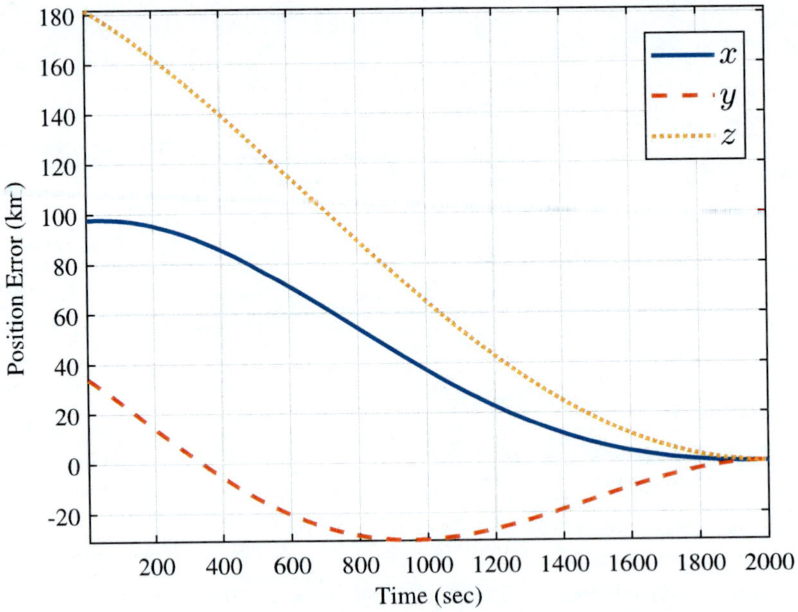

Fig. 7.16 Position error history ($\rho_i = 1$ km, $\rho_f = 100$ km, $e = 0.2$)

Table 7.3 MPSP terminal state error $\left[||d\mathbf{Y}_N|| \leq 5.5 \times 10^{-4}\right]$

Error in states	Initial LQR guess	Itr#1	Itr#2	Itr#3	Itr#4
x	2.36×10^{-1}	-9.40×10^{-3}	-1.70×10^{-3}	-3.11×10^{-4}	-2.59×10^{-5}
$\dot{x}$	2.79×10^{-4}	-8.11×10^{-6}	-3.33×10^{-6}	-7.52×10^{-7}	-8.29×10^{-8}
y	-2.33×10^{-2}	-6.14×10^{-4}	-1.87×10^{-3}	-4.65×10^{-4}	-5.61×10^{-5}
$\dot{y}$	-1.55×10^{-4}	1.03×10^{-5}	-1.30×10^{-6}	-7.56×10^{-7}	-1.45×10^{-7}
z	-7.38×10^{-3}	-2.58×10^{-6}	-5.04×10^{-8}	-9.21×10^{-9}	-5.72×10^{-10}
$\dot{z}$	-4.46×10^{-6}	-6.20×10^{-9}	-2.80×10^{-10}	-4.72×10^{-11}	-3.37×10^{-12}
Norm	2.37×10^{-1}	9.42×10^{-3}	2.53×10^{-3}	5.60×10^{-4}	6.17×10^{-5}

Table 7.4 G-MPSP terminal state error $\left[||d\mathbf{Y}_N|| \leq 5.5 \times 10^{-4}\right]$

Error in states	Initial LQR guess	Itr#1	Itr#2	Itr#3
x	2.36×10^{-1}	-9.05×10^{-3}	-1.72×10^{-3}	-3.10×10^{-4}
$\dot{x}$	2.79×10^{-4}	-8.00×10^{-6}	-3.38×10^{-6}	-7.44×10^{-7}
y	-2.33×10^{-2}	-9.24×10^{-4}	-1.89×10^{-3}	-4.57×10^{-4}
$\dot{y}$	-1.55×10^{-4}	9.85×10^{-6}	-1.39×10^{-6}	-7.49×10^{-7}
z	-7.38×10^{-3}	1.89×10^{-6}	-5.89×10^{-8}	-9.44×10^{-9}
$\dot{z}$	-4.46×10^{-6}	-3.37×10^{-9}	-2.84×10^{-10}	-4.75×10^{-11}
Norm	2.37×10^{-1}	9.09×10^{-3}	2.55×10^{-3}	5.50×10^{-4}

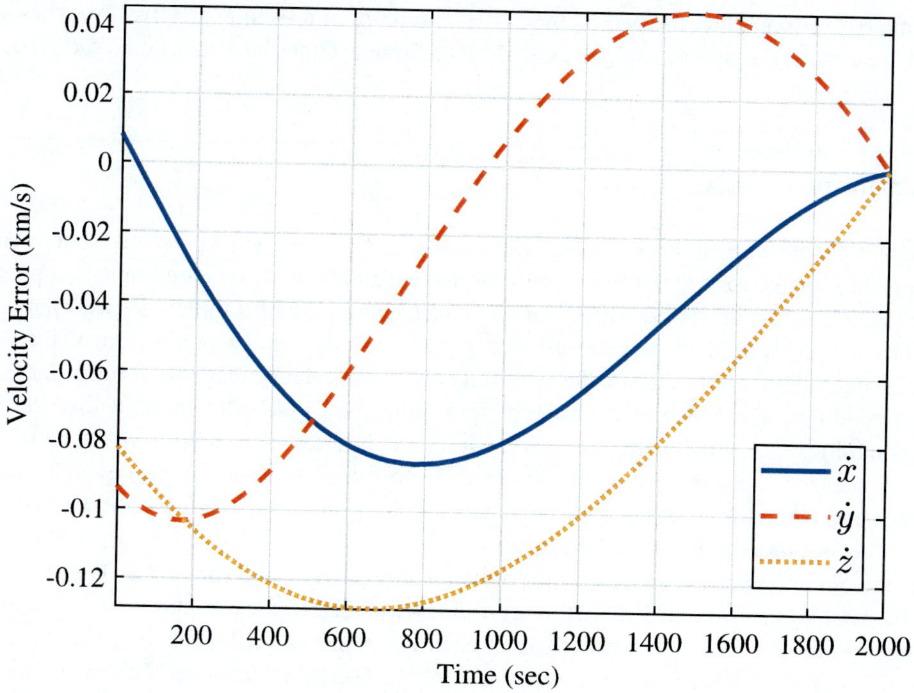

Fig. 7.17 Velocity error history ($\rho_i = 1$ km, $\rho_f = 100$ km, $e = 0.2$)

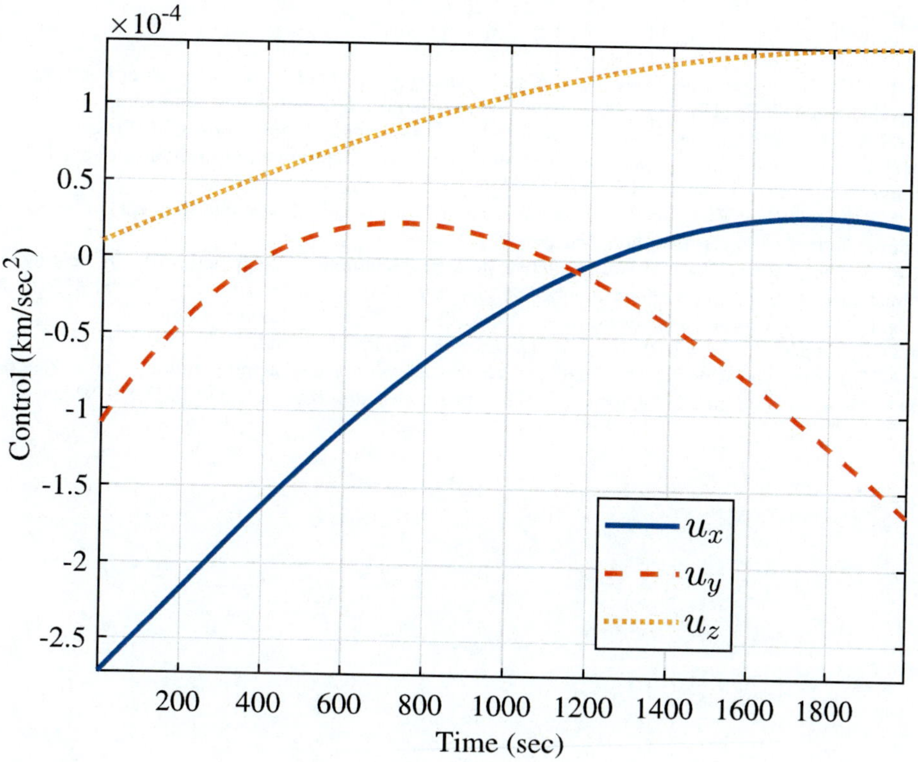

Fig. 7.18 Control history for DS ($\rho_i = 1$ km, $\rho_f = 100$ km, $e = 0.2$)

In summary, we found that the two methods perform more or less equivalently. If one has a faster processor, we would recommend using the G-MPSP technique. Otherwise, the choice should be to use MPSP.

7.7 Summary

In this chapter, the theoretical details of MPSP and G-MPSP methods were presented first. Numerical simulation studies were carried out to demonstrate the capability, and compared to the outputs of the finite-time SDRE technique as well. Three cases with increasing order of complexities were handpicked, and simulations were carried out successfully. A comparison study was also presented with the F-SDRE technique under elliptic orbit with small eccentricity and small relative distance. It turns out that even under this condition, the MPSP and G-MPSP techniques perform better in achieving the desired terminal conditions with much better accuracy. Moreover, it turns out that both MPSP and G-MPSP methods are successful in synthesizing the guidance history even for severely penalizing scenarios of formation flying with elliptic orbits with large eccentricities and large relative distance and that too under the influence of J_2 gravitational influence.

About the computational complexity performance comparison, we found that the two methods perform more or less equivalently. G-MPSP converges with slightly lesser number of iterations than MPSP, whereas the number of computations in each iteration is a bit higher than MPSP. Hence, if one has a faster processor, we would recommend using the G-MPSP technique. Otherwise, the choice should be MPSP.

References

1. Keller, H.B. 2018. *Numerical methods for two-point boundary-value problems*. Courier Dover Publications.
2. Bryson, A.E., and Y.C. Ho. 1975. *Applied optimal control*. Hemisphere Publishing Corporation.
3. Rossiter, J.A. 2003. *Model based predictive control: A practical approach*. New York: CRC Press.
4. Werbos, P.J. 1992. *Approximate dynamic programming for real-time control and neural modeling*, ed. D.A. White, D.A. Sofge. New York: Van Nostrand Reinhold.
5. McHenry, R.L., A.D. Long, B. Cockrell, J. Thibodeau, and T.J. Brand. 1979. Space shuttle ascent guidance, navigation, and control. *Journal of the Astronautical Sciences* 27 (1): 1–38.
6. Padhi, R., and M. Kothari. 2009. Model predictive static programming: A computationally efficient technique for suboptimal control design. *International Journal of Innovative Computing, Information and Control* 5 (2): 399–411.
7. Kreyszig, E. 2009. *Advanced engineering mathematics*, 10th ed. Technical Report, Wiley.
8. Ewing, G.M. 1985. *Calculus of variations with applications*. Courier Corporation.
9. Maity, A., H.B. Oza, and R. Padhi. 2014. Generalized model predictive static programming and angle-constrained guidance of air-to-ground missiles. *Journal of Guidance, Control, and Dynamics* 37 (6): 1897–1913.

8 Performance Comparison

By now, it must be obvious to the reader that the major goal for this book is to present various modern optimal control and adaptive guidance design techniques that are best suited in application to the satellite formation flying problems. Namely, the following design techniques have been discussed in detail in various preceding chapters: (i) Infinite-time linear quadratic regulator (LQR), (ii) Infinite-time state-dependent Riccati Equation (I-SDRE), (iii) Adaptive LQR, (iv) Adaptive dynamic inversion (DI) (v) Finite-time linear quadratic regulator (F-LQR), (vi) Finite-time state-dependent Riccati Equation (F-SDRE), and (vii) both discrete and continuous versions of model predictive static programming (MPSP). It is apparent, a curious reader would definitely want a comparative study to figure out the best performing method for the problem in hand. This chapter is aimed at to bring out the performance aspects of various methods presented in the earlier chapters.

First, for better clarity ideally, only two techniques need to be compared together at a time. However, with this, one can notice that there are ${}^{7}C_2 = 21$ combinations. Including all these 21 combination of results is a bit too much and carries the danger of confusing the reader as well. Hence, to limit the volume and yet include a meaningful discussion, the possible combinations are restricted to only 'similar' techniques and categorized into two sets:

(i) Adaptive LQR and Adaptive DI
(ii) F-SDRE and MPSP.

The first set contains techniques from the regulator and asymptotic stability theory (where the error is driven to zero in infinite-time), whereas the second set contains finite-time optimal control theory (where the error is driven to zero at a pre-selected final time). Note that the adaptive versions of LQR and DI are included here since these give better results in presence of elliptic orbits, larger relative distance, and J_2 perturbation, as compared to their non-adaptive versions (see Chapters 4 and 5 for details).

It turns out that the second category of techniques is not suitable for adaptive control augmentation as the adaptation theory is based on asymptotic stability theory, whereas both F-SDRE and MPSP are based on finite-time theory. However, it is interesting to observe that, even without having adaptive control augmentation, these techniques can handle the punishing conditions such as chief satellite in elliptic orbit, larger relative distance, and J_2 perturbation effectively because of the very nature of the problem formulation and associated computations. Therefore, the results of *these techniques can be compared with the results obtained from the adaptive LQR and adaptive DI techniques under the same simulation*

S. Mathavaraj and R. Padhi, *Satellite Formation Flying*,
https://doi.org/10.1007/978-981-15-9631-5_8

environment. At this juncture, it can be stressed that the effective use of SDRE technique demands that one uses the appropriate SDC form to derive maximum benefit out of it. From the observations made in Chapter 6, SDC_2 form turns out to be better among the available choices and hence only that is selected. The MPSP technique, on the other hand, has two choices, namely the discrete-time MPSP and continuous-time G-MPSP. A comparison of performance using MPSP and G-MPSP has been presented in Chapter 7. It turns out that both result in 'similar performance'; however, G-MPSP converges in slightly lesser number of iterations, but with more computation per iteration. Hence, only the discrete-time MPSP technique is selected here for the performance comparison purpose.

Without loss of generality, the performance of these techniques is demonstrated when the chief satellite is in eccentric orbit with large eccentricity and for large desired relative distance considering J_2 perturbation, where the linearization assumption is not valid. This scenario is considered because it is the most punishing case among all cases (other cases can be considered as special cases of this scenario). The initial condition and the desired final condition at a fixed final time of 2000 s for this simulation study are selected as in Table 2.4 (see Chapter 2).

In the following sections, details of these comparison studies are discussed. Based on these comparisons, final set of conclusions are made and recommendations are provided in the next chapter. It is noteworthy to realize that even though the first set of comparison studies (i.e., Adaptive LQR and Adaptive DI) are based on infinite-time theory, for the sake of consistency those cases are also simulated only upto 2000 s.

8.1 Comparison Studies: Adaptive LQR and Adaptive DI

In this section, a comparison study among Adaptive LQR and Adaptive DI is carried out. The trajectories of the deputy satellite in Hill's frame using the two different techniques (guidance schemes) are shown in Fig. 8.1. The relative position and velocity error histories of the above-said techniques are plotted in Figs. 8.2 and 8.3, respectively, which show that the state errors progressively reduce along the trajectory

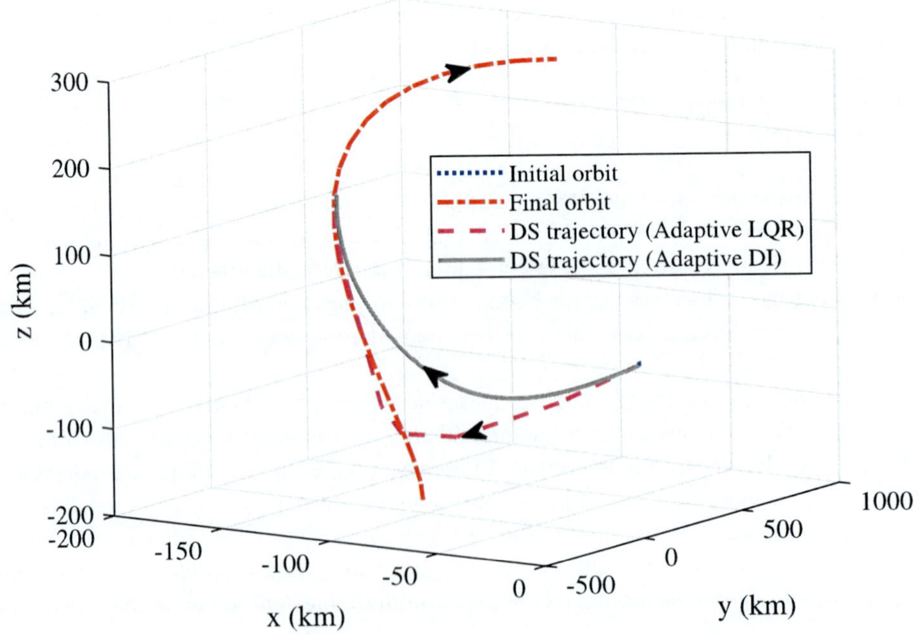

Fig. 8.1 Deputy satellite (DS) trajectory ($\rho_i = 1$ km, $\rho_f = 100$ km, $e = 0.2$)

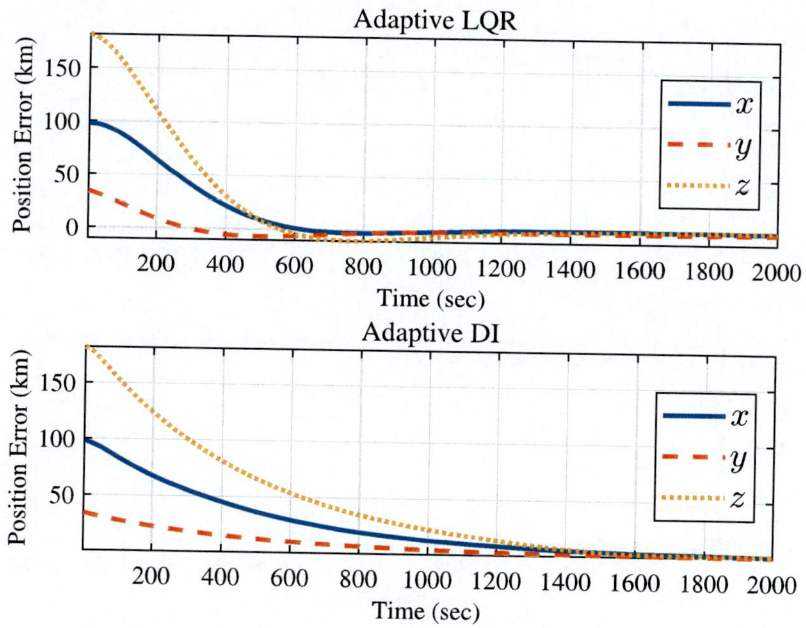

Fig. 8.2 Position error history ($\rho_i = 1$ km, $\rho_f = 100$ km, $e = 0.2$)

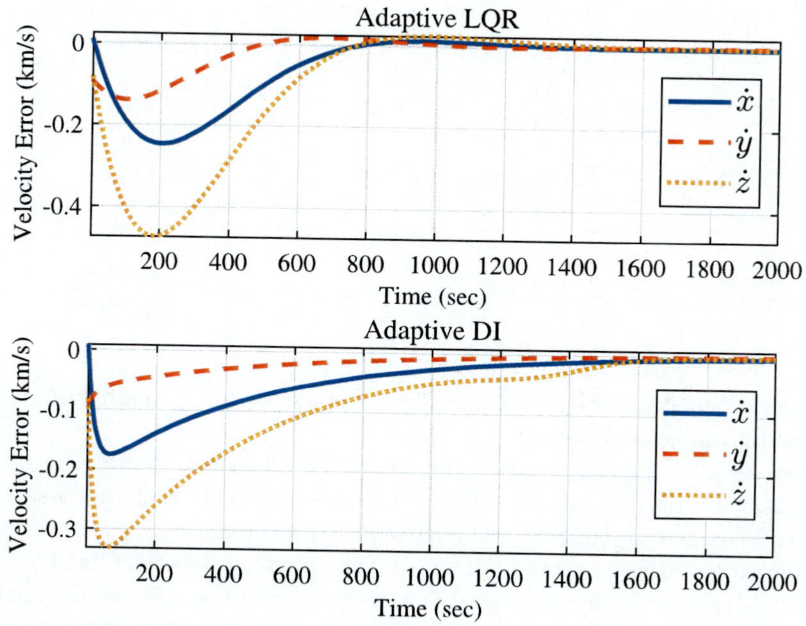

Fig. 8.3 Velocity error history ($\rho_i = 1$ km, $\rho_f = 100$ km, $e = 0.2$)

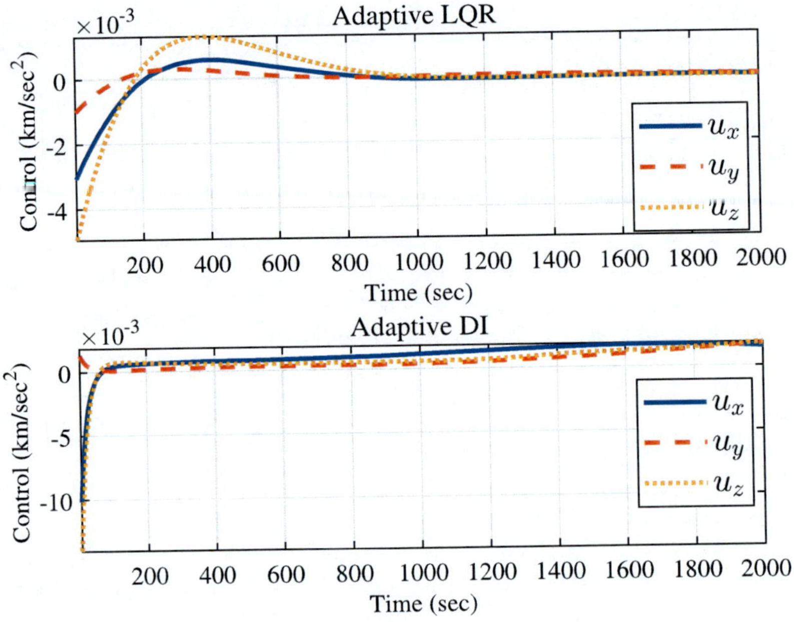

Fig. 8.4 Control history for deputy satellite ($\rho_i = 1$ km, $\rho_f = 100$ km, $e = 0.2$)

Table 8.1 Terminal state error ($\rho_i = 1$ km, $\rho_f = 100$ km, $e = 0.2$)

State error	Adaptive LQR	Adaptive DI
Δx (km)	-4.17×10^{-1}	1.20×10^{0}
$\Delta \dot{x}$ (km/s)	-8.64×10^{-4}	-3.68×10^{-3}
Δy (km)	-8.51×10^{-1}	1.89×10^{-2}
$\Delta \dot{y}$ (km/s)	-1.70×10^{-3}	-4.44×10^{-4}
Δz (km)	7.71×10^{-1}	1.34×10^{0}
$\Delta \dot{z}$ (km/s)	2.82×10^{-4}	-2.15×10^{-3}
Norm	1.22×10^{0}	1.81×10^{0}

to achieve the desired condition. Figure 8.4 gives the generated necessary control histories of Adaptive LQR and Adaptive DI to achieve the desired states.

First, as seen clearly in Fig. 8.1, the Adaptive LQR drives the states to progressively achieve the desired states at final time although the chief satellite is in eccentric orbit with a sufficiently large ($e = 0.2$) eccentricity and for large desired relative distance considering J_2 perturbation. Note that Adaptive LQR has been derived using LQR as baseline controller. However, the terminal accuracy achieved by Adaptive LQR is superior than LQR (refer to Section 4.4.2). Implementationwise, the logic is simple and does not involve onboard computational burden. The key benefit of this idea is that the deputy satellite can continue to implement LQR controller neglecting the nonlinear plant and J_2 perturbation model. However, augmenting it with the proposed neural network-based adaptive nonlinear optimal controller ensures that it indirectly recovers the neglected dynamics automatically and accounts for it in the revised controller, thereby leading to better performance.

It can be observed from the same Fig. 8.1 that the Adaptive DI achieves the desired states by synthesizing an adaptive control to compensate for the unknown J_2 disturbances using dynamic inversion as baseline controller. The Adaptive DI controller is clearly superior to nominal DI controller in terms of terminal accuracy achieved (refer to Section 5.4.2). Note that the augmented neural network captures the unknown disturbances which make the DI controller robust to parameter inaccuracy and modeling errors. The achieved errors of final states of Adaptive LQR and Adaptive DI are tabulated in Table 8.1. It can be mentioned here that the terminal accuracy of both techniques are of same order in the overall norm sense.

However, as evident in Fig. 8.4, the control histories of the Adaptive LQR show lesser transient behavior as compared to the Adaptive DI. Moreover, in Adaptive LQR approach the control histories approach to zero and stay at zero much before the final time, which are very desirable features. Hence, if designer has to choose one out of these two, the Adaptive LQR technique is recommended.

8.2 Comparison Studies: F-SDRE and MPSP

In this section, a comparison study between the SDC_2-based F-SDRE and MPSP techniques is carried out. The resulting relative trajectories of deputy satellite in Hill's frame are shown in Fig. 8.5. As it is seen in Fig. 8.5, both guidance schemes are capable of driving the deputy satellite to the desired relative states even when the chief satellite is in eccentric orbit with large eccentricity, for large desired relative distance and considering J_2 perturbation. This is because of the inherent nonlinear nature of these techniques. The position and velocity error histories are plotted in Figs. 8.6 and 8.7, respectively. These figures show that the state errors reduce along the trajectory to achieve the desired condition. Figure 8.8 illustrates the control histories required for achieving the desired state values. The control history profiles of these two techniques do not look similar because they operate on a slightly different nonlinear relative dynamics, i.e., F-SDRE operates on the approximate SDC_2 model, whereas MPSP operates on the complete nonlinear model considering J_2 perturbation. The computed final state errors for the cases due to F-SDRE and MPSP techniques are tabulated in Table 8.2.

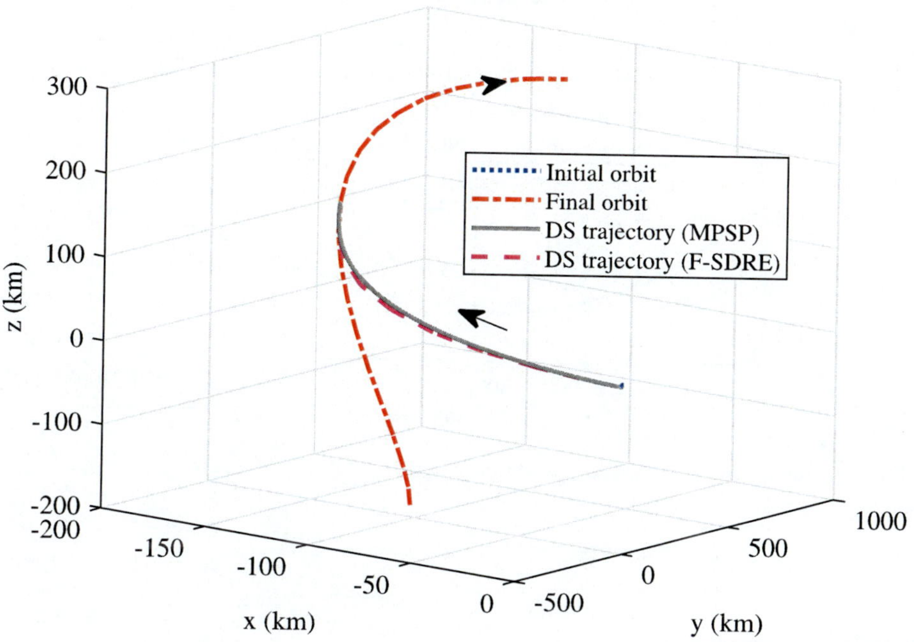

Fig. 8.5 Deputy satellite (DS) trajectory ($\rho_i = 1$ km, $\rho_f = 100$ km, $e = 0.2$)

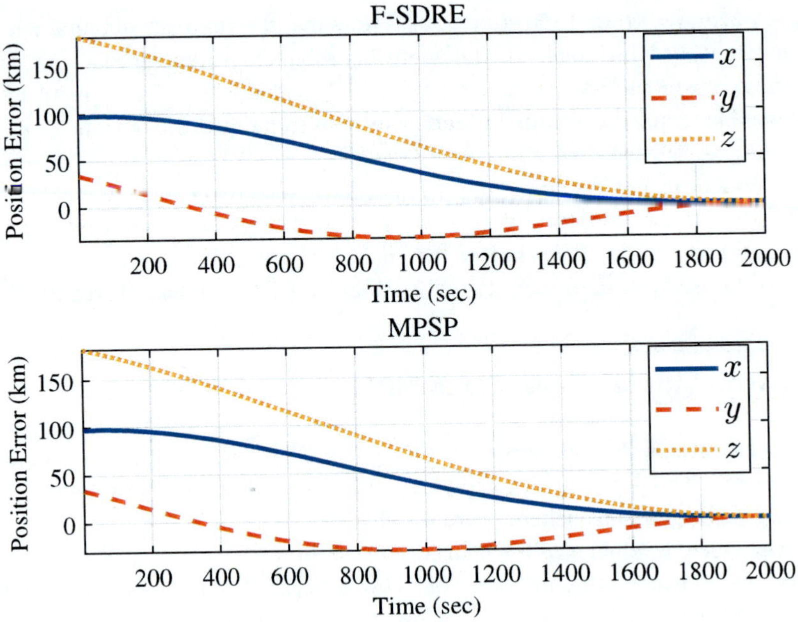

Fig. 8.6 Position error history ($\rho_i = 1$ km, $\rho_f = 100$ km, $e = 0.2$)

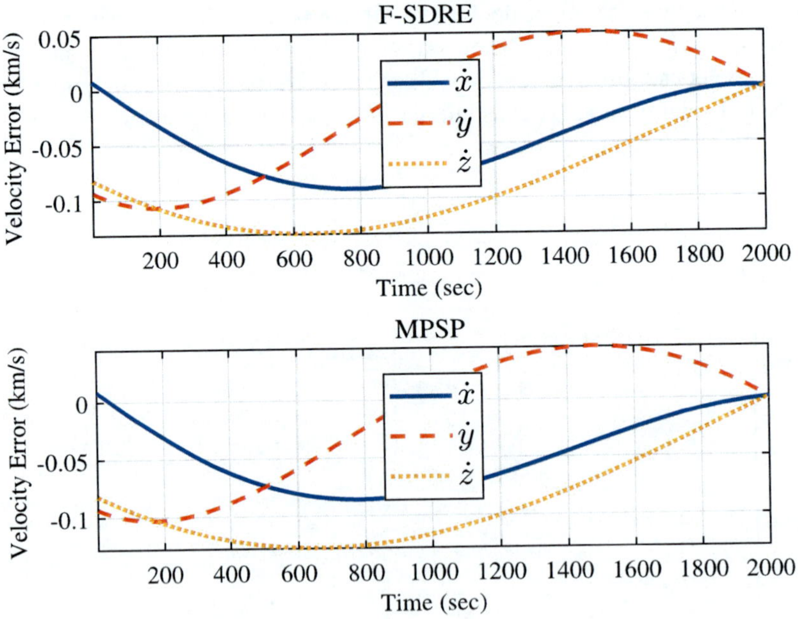

Fig. 8.7 Velocity error history ($\rho_i = 1$ km, $\rho_f = 100$ km, $e = 0.2$)

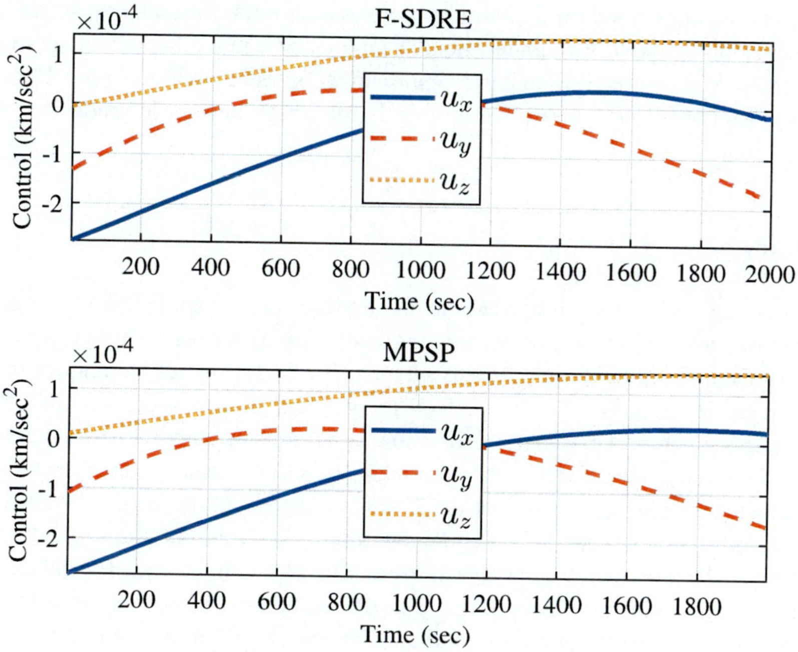

Fig. 8.8 Control history for deputy satellite ($\rho_i = 1\,\text{km}$, $\rho_f = 100\,\text{km}$, $e = 0.2$)

Table 8.2 Terminal state error ($\rho_i = 1\,\text{km}$, $\rho_f = 100\,\text{km}$, $e = 0.2$)

State error	F-SDRE	MPSP
Δx (km)	-1.28×10^{-2}	-1.32×10^{-4}
$\Delta \dot{x}$ (km/s)	-8.03×10^{-5}	-3.02×10^{-7}
Δy (km)	2.08×10^{-1}	-1.55×10^{-4}
$\Delta \dot{y}$ (km/s)	-1.66×10^{-5}	-2.49×10^{-7}
Δz (km)	1.53×10^{-1}	5.92×10^{-7}
$\Delta \dot{z}$ (km/s)	-6.16×10^{-5}	1.20×10^{-9}
Norm	2.59×10^{-1}	2.04×10^{-4}

From a simple observation of Tables 8.1 and 8.2, it is obvious that the achieved terminal accuracy by F-SDRE is better than the Adaptive LQR and Adaptive DI. This is attributed to the finite-time formulation of the SDRE solution, and it is based on the SDC_2 approximation of the system model that retains the nonlinearity of the SFF problem to the maximum extent possible. It should be mentioned here that the SDC_2 does not involve the J_2 perturbation effect in the formulation, as deriving an SDC form including this perturbation force is a substantially challenging task.

Now it can be observed from Table 8.2, and the terminal error achieved by MPSP is much lesser than even the SDC_2- based F-SDRE technique. This result can be attributed to the fact that (i) MPSP is also a finite-time formulation with terminal hard constraint and (ii) the method approximates the nonlinear nature of the problem, inclusive of J_2 perturbation, better. It turns out that the terminal accuracy achieved by MPSP is superior than other approaches for various problem scenario like non-zero eccentric chief satellite, large baseline separation, J_2 perturbation, and so on (refer to Section 7.3). Hence, the MPSP method is strongly recommended for handling complex satellite formation flying scenarios. However, even though the technique leads to high terminal accuracy and a few other advantages (see Section 7 for details), a small caveat is that it is an iterative numerical optimization technique. Hence, the MPSP method requires faster onboard processors. Fortunately, such high-end space-grade processors are now available [1].

8.3 Summary

In this chapter, a comparison study is carried out among four 'prominent techniques' discussed in this book. Justification of selecting these four techniques, and grouping those into two sets of two techniques each, has been given at the beginning of this chapter. To be fair to all techniques, the simulation environment is made identical for this comparison study, which encompasses various parameters such as (i) the chief satellite's orbital parameters, (ii) initial conditions of both the satellites, (iii) the desired final condition of the deputy satellite, (iv) the system dynamics and associated parameters, (v) the final-time selection, and (vi) time step of simulation. The advantages and disadvantages of each technique have been discussed. Among these four techniques, it turns out that *the finite-time formulations are superior as compared to the infinite-time formulations*. Out of two finite-time formulations, the MPSP guidance scheme outperforms the SDC_2-based F-SDRE technique. Moreover, since the MPSP is free from the additional difficulty of selecting a good SDC form, one can implement it in a straightforward manner with relative ease. However, one should be aware that, unlike the other three techniques, it is an *iterative technique*. Hence, one has to make a judicious call about it's usage even though it has fast convergence properties. From these comparison studies, as well as several other comparison studies done throughout various preceding chapters, the final set of conclusions are made and recommendations are provided in the next chapter.

Reference

1. Lovelly, T.M., and A.D. George. 2017. Comparative analysis of present and future space-grade processors with device metrics. *Journal of Aerospace Information Systems* 14 (3): 184–197.

9 Conclusion

An emerging trend across the globe is to have missions involving many small, distributed, and largely inexpensive satellites flying in formation to achieve a common objective. The advantages of satellite formation flying can be summarized as (i) higher redundancy and improved fault tolerance, (ii) on-orbit reconfiguration within the formation (which offers multi-mission capability and design flexibility), (iii) lower individual launch mass in case of small satellite missions, typically leading to reduced launch cost and increased launch flexibility, and (iv) minimal financial loss in case of failures both during launch as well as operation. Satellite formation flying enables new application areas such as spar antenna arrays for remote sensing, distributed sensing for solar and extra-terrestrial observatories, interferometry synthetic aperture radar, and many more. Also a variety of interesting and useful applications are possible in satellite formation flying, both as a substitute to single large satellite missions as well as the ones unique to formation flying missions.

In order to ensure the formation flying of satellites, the basic understanding of the space dynamics is very essential. Moreover, the advanced guidance schemes discussed in this book rely on model-based control. In order to familiarize the readers, an overview of the Keplerian two-body problem, where two celestial bodies keep moving under the gravitational influence of each other, is presented. As a special case, the equation of motion for a small body orbiting a large body that reflects the motion of orbiting satellites around the Earth has been derived. Subsequently, the relative motion between two orbiting satellites in Hill's frame has been presented, which leads to the well-known Clohessy–Wiltshire (C-W) equations. As a special case, when the chief satellite is in a circular orbit and the relative distance is small, a very useful linearized Hill's equation comes into picture. Details of these equations are given in Chapter 2 of this book.

The reader might have been surprised that in the entire book the spacecraft attitude dynamics has not been considered, even though controlling a satellite for mission objectives typically involves both orbit and attitude control. This can be justified because, in case of satellite formation flying applications, one can decouple the translational and rotational dynamics because of the following reasons: (i) the force due to reaction control thruster firing in order to achieve the desired attitude does not significantly affect the orbital motion, and (ii) due to the lack of atmosphere, the change in the spacecraft orientation do not affect the linear momentum of the spacecraft.

Nevertheless, the primary focus of this book is to identify and present the advanced guidance schemes to address the orbit correction problem in order to ensure the desired formation flying. The design techniques discussed in detail in several chapters include (i) Infinite-time linear quadratic

S. Mathavaraj and R. Padhi, *Satellite Formation Flying*,
https://doi.org/10.1007/978-981-15-9631-5_9

regulator (LQR), (ii) Infinite-time state-dependent Riccati Equation (I-SDRE)—SDC_1 and SDC_2, (iii) Adaptive LQR, (iv) Adaptive dynamic inversion, (v) Finite-time LQR, (vi) Finite-time SDRE—SDC_1 and SDC_2, and (vii) both discrete and continuous versions of model predictive static programming (MPSP). A comprehensive application scenarios and the methods presented in this book are given in a matrix form in Table 9.1, which can be referred to as and when required.

In infinite-time linear quadratic regulator (LQR) theory discussed in Chapter 3, the optimal control has been derived based on linear system dynamics while minimizing the quadratic cost function. Even though the results of the infinite-time LQR approach are seemingly easier to understand and implement and can lead to acceptable results with appropriate tuning, it is very important to understand that such infinite-time approaches are *not strongly recommended* for satellite formation flying problems in general. A right problem formulation for satellite formation flying should ensure that the relative desired position and velocity vectors are achieved at a '*particular time*' (not earlier, not later). Hence, the problem formulation should ideally be carried out under the '*finite-time*' optimal control paradigm instead. To address such finite-time terminal constraints problems, finite-time LQR approach based on state transition matrix approach is presented in Chapter 6.

Implementation of infinite-time/finite-time LQR is relatively simple, and there is not much computational burden for online implementation. The technique shows acceptable terminal accuracy when the chief satellite is in circular orbit and the desired relative distance is small. This is because the LQR theory is based on linearized dynamics, which is valid under these conditions (refer to Section 3.5.1). However, for other non-conductive scenarios, the method does not perform well in the sense that it leads to a low terminal accuracy. Hence, the LQR technique can be recommended only when the chief satellite is in circular orbit and the desired relative distance is small. If one has to choose between the two, the finite-time LQR is preferred as it leads to better results for the reason mentioned earlier.

In order to address the limiting issues of the LQR-based techniques discussed above and to account for the nonlinear relative dynamics, the nonlinear infinite-time state-dependent Riccati Equation (SDRE) guidance scheme is proposed in Chapter 3. Note that the SDRE technique is an intuitive extension of the LQR design under specific assumptions, the details of which are presented in Section 3.3. The key idea in modifying LQR to suit the problem is to describe the system dynamics in linear-looking state-dependent coefficient (SDC) form. Next, the control solution, which serves as the guidance command, is derived by repeatedly solving the corresponding Riccati equation online at every grid point of time. However, the non-uniqueness of the SDC parameterization of the system dynamics poses a major challenge in the successful implementation of the SDRE technique. It may also restrict the validity domain of the resulting controller. In this book, two SDC formulations, namely SDC_1 and SDC_2 parameterizations, for the nonlinear C-W equation are presented. Both of these represent the nonlinear relative dynamics between the two satellites. In the SDC_1 formulation, the C-W equation is simplified using the assumption that the chief satellite is in the circular orbit. Therefore, the SDC_1 representation in Eq. (3.23) only caters to the circular reference orbit scenario of satellite formation flying. When this assumption is not true, it leads to the erroneous results and hence it is recommended only when the chief satellite is in circular orbit (refer to Table 9.1). The second approach using the SDC_2 formulation is a better representation of the nonlinear scenario and hence (i) the method is suitable even if the chief satellite is in a mild to moderate elliptical orbit, and (ii) the relative distance between the chief and deputy satellite need not be small.

Next, an Adaptive LQR is presented in Chapter 4, which has been derived using LQR as baseline controller and augmented with neural network concepts in order to take care of nonlinearity of the problem. As demonstrated in Section 4.4, the terminal accuracy achieved by Adaptive LQR is superior to LQR for scenarios like small eccentric orbit of chief satellite, large desired relative formation, J_2 perturbation acting on the satellite, etc., as portrayed in Table 9.1. Furthermore, implementationwise, the logic is fairly simple.

Table 9.1 Guidance techniques and their applicability scenarios

Techniques	Circular orbit and small relative distance	Circular orbit and large relative distance	Eccentric orbit and small relative distance	Eccentric orbit and large relative distance	Highly eccentric orbit and large relative distance	J_2 perturbation	Closed form
Infinite-time LQR	✓	×	×	×	×	×	✓
Finite-time LQR	✓	×	×	×	×	×	✓
Infinite-time SDC_1	✓	✓	×	×	×	×	✓
Infinite-time SDC_2	✓	✓	✓	✓	×	×	✓
Adaptive LQR	✓	✓	✓	✓	×	✓	✓
Adaptive DI	✓	✓	✓	✓	×	✓	✓
Finite-time SDC_1	✓	✓	×	×	×	×	✓
Infinite-time SDC_2	✓	✓	✓	✓	×	×	✓
MPSP	✓	✓	✓	✓	✓	✓	×
G-MPSP	✓	✓	✓	✓	✓	✓	×

✓ Applicable
× Not Applicable

Yet another Adaptive dynamic inversion guidance is presented for satellite formation flying problem in the next chapter. This method synthesize an adaptive control to compensate for the unknown disturbances using dynamic inversion as the baseline controller. For obvious reasons (see Chapter 5 for details), it turns out that the performance of the Adaptive DI controller is clearly superior to nominal DI controller in terms of terminal accuracy achieved for many practical scenarios like small eccentric orbit of chief satellite, large desired relative formation, J_2 perturbation acting on the satellite (refer to Section 5.4). In general, the augmented neural network captures the unknown disturbances which make the dynamic inversion controller robust to parameter inaccuracy as well as the modeling errors.

In Chapter 6, the state transition matrix-based finite-time SDRE approach is presented. Note that unlike infinite-time approach, the finite-time method attempts to achieve the final states as 'hard constraints' at the final time and hence leads to better results. As seen in Sections 6.3 and 8.2, the SDC_2 formulation is significantly less at the achieved final sate error as it retains the nonlinearity of the SFF problem to the maximum extent possible. In principle, the terminal accuracy of the technique depends on how well the choice of SDC form $A(\mathbf{x})\,\mathbf{x}$ captures the function $f(\mathbf{x})$ in the system dynamics.

Next, the MPSP approach is presented in Chapter 7. This approach computes the control history by formulating a nonlinear optimal control problem with hard boundary condition at a pre-selected final time and solving it in a computationally efficient manner. As demonstrated in Chapter 8, as compared to other guidance techniques, the MPSP approach is far superior in achieving terminal accuracy, especially when the satellite formation flying scenarios are more demanding. Moreover, the method is quite generic and is applicable to various problem scenarios such as elliptic orbits with large eccentricity, large desired separation distance between the two satellites, J_2 perturbation, and so on (see Table 9.1). On the other hand, it is a 'computational guidance' and iterations are necessary before convergence. However, it is a computationally efficient technique with good convergence properties, i.e., it converges with very less number of iterations in general and the computations necessary for each iteration is quite minimal. Fortunately, high-performance space-grade processors are now available [1], which can facilitate its onboard implementation.

One has the choice to choose either discrete-time MPSP or generalized MPSP based on the following arguments. Similar to MPSP, the G-MPSP approach also includes the computation of a time-varying weighting matrix $W(t)$, iterative update of control $\boldsymbol{u}(t)$ history. For achieving the desired output within the specified tolerance, it is observed that G-MPSP method converges within lesser number of iterations than the number of iterations required in MPSP (refer to Section 7.6.3 for details). This is mainly because it is relatively much easier to use higher order numerical methods in the G-MPSP technique owing to the continuous-time formulation. However, because of this, the G-MPSP method also requires more computations per iteration. Since both lead to more or less same performance in accuracy, if one has a faster processor, we would recommend using the G-MPSP technique. Otherwise, the choice should be to use MPSP.

For the convenience of the reader, these observations are summarized in Table 9.1. The reader/user can easily weigh the pros and cons in selecting a suitable technique for the scenario that he/she is interested in.

Once the technique has been chosen by the user, the desired guidance command will be generated based on the satellite formation flying scenario considered. In general, this guidance command will be realized using the discrete mode operation of reaction control system thrusters as discussed in the Section 2.5. Further details on this topic can be found in [2].

References

[1] Lovelly, T.M., and A.D. George. 2017. Comparative analysis of present and future space-grade processors with device metrics. *Journal of Aerospace Information Systems* 14 (3): 184–197.

[2] Buck, N.V. 1996. Minimum vibration maneuvers using input shaping and pulse-width, pulse-frequency modulated thruster control. Technical Report, Naval Postgraduate School Monterey, CA.

A Program Files Documentation

The well-documented MATLAB codes, from which the results appearing in this book have been generated, are provided in the publisher's website for the benefit of the readers. In this chapter, a brief summary about each program folder is discussed so that reader can easily get acquainted while executing the files in it. The platform for all the source files is in MATLAB. Namely, the program files for the following design techniques have been provided: (i) Infinite-time linear quadratic regulator (LQR), (ii) Infinite-time state-dependent Riccati Equation (SDRE), (iii) Adaptive LQR, (iv) Adaptive dynamic inversion (DI) (v) Finite-time LQR, (vi) Finite-time SDRE, and (vii) both discrete and continuous versions—Model predictive static programming (MPSP). These techniques are documented in different folders in the same way the chapters of this book are classified. Interested reader can simulate all the results presented in this book by feeding the appropriate initial and desired condition discussed in Section 2.6.

A.1 Infinite-Time: LQR and SDRE

The script file 'MAIN.m' is the start source file for executing this program. This folder generates the satellite formation trajectory using the infinite-time LQR and SDRE guidance methods discussed in Chapter 3. Note that the program files in this folder comprise both SDC_1 and SDC_2 formulations of SDRE philosophy.

A.2 Adaptive LQR

This folder generates the satellite formation trajectory using the adaptive LQR technique presented in Chapter 4. Both neural networks NN_1 and NN_2 are coded with appropriate weight update rule. Moreover, the baseline LQR controller is programmed to augment with these neural networks as per the adaptive philosophy. The script file 'MAIN.m' is the start source file for executing this program.

S. Mathavaraj and R. Padhi, *Satellite Formation Flying*,
https://doi.org/10.1007/978-981-15-9631-5

A.3 Adaptive DI

This folder generates the satellite formation trajectory using the adaptive DI technique discussed in Chapter 5. The baseline DI controller is programmed to augment with the neural network so that enhanced performance is achieved at the terminal condition. The script file 'MAIN.m' is the start source file for executing this program.

A.4 Finite-Time: LQR and SDRE

The script file 'MAIN.m' is the start source file for executing this program. This folder generates the satellite formation trajectory using the finite-time LQR and SDRE guidance methods discussed in Chapter 6. Note that the program files in this folder comprise both SDC_1 and SDC_2 formulations of SDRE philosophy.

A.5 MPSP

This folder contains two subfolders named 'D-MPSP' and 'G-MPSP'. The program files in these subfolders generate the satellite formation trajectory using the model predictive static programming guidance methods—discrete MPSP and generalized MPSP discussed in Chapter 7. The script file 'MAIN.m' in each subfolder is the start source file for executing the corresponding programs.

Index

S. Mathavaraj and R. Padhi, *Satellite Formation Flying*,
https://doi.org/10.1007/978-981-15-9631-5